Symmetry Application in Motor Control in Sports and Rehabilitation

Guest Editors

Arthur de Sá Ferreira
Fabio Vieira dos Anjos

MDPI

Basel • Beijing • Wuhan • Barcelona • Belgrade • Novi Sad • Cluj • Manchester

Guest Editors
Arthur de Sá Ferreira
Centro Universitário
Augusto Motta
Rio de Janeiro
Brazil

Fabio Vieira dos Anjos
Centro Universitário
Augusto Motta
Rio de Janeiro
Brazil

Editorial Office
MDPI AG
Grosspeteranlage 5
4052 Basel, Switzerland

This is a reprint of the Special Issue, published open access by the journal *Symmetry* (ISSN 2073-8994), freely accessible at: https://www.mdpi.com/journal/symmetry/special_issues/Symmetry_Application_Motor_Control_Sports_Rehabilitation.

For citation purposes, cite each article independently as indicated on the article page online and as indicated below:

Lastname, A.A.; Lastname, B.B. Article Title. *Journal Name* **Year**, *Volume Number*, Page Range.

ISBN 978-3-7258-6744-8 (Hbk)
ISBN 978-3-7258-6745-5 (PDF)
https://doi.org/10.3390/books978-3-7258-6745-5

Contents

symmetry

MDPI

Editorial

Applying Symmetry to Motor Control in Sports and Rehabilitation

Arthur de Sá Ferreira * and Fabio Vieira dos Anjos *

Postgraduate Program in Rehabilitation Science (PPGCR/UNISUAM), Centro Universitário Augusto Motta, Rio de Janeiro 21041-020, Brazil
* Correspondence: arthurde@souunisuam.com.br (A.d.S.F.); fabioanjos@souunisuam.com.br (F.V.d.A.)

Symmetry is a core yet controversial concept in the health sciences, spanning multiple disciplines from anatomy and biomechanics to pathophysiology, with important clinical implications for both diagnostic and therapeutic practices [1]. In human movement science, symmetry has traditionally been regarded as an indicator of optimal function [2]; however, growing evidence suggests that asymmetry may represent not only impairment, but also adaptation, compensation, or task-specific specialization.

In recent years, there has been renewed interest in investigating symmetry and asymmetry in biomechanical and functional aspects of individuals with neurological and musculoskeletal conditions, as well as in athletic and physically active populations. Although body (as)symmetry is often easily recognized using subjective observation, its objective quantification (whether structural or functional) remains methodologically challenging. Moreover, the effects of symmetry-related factors—such as posture alignment [3], sidedness, bodyweight distribution [4], muscle activation on postural control [5], activities of daily living [5], injury risk [6], and sports performance [7]—are not yet fully understood. Clarifying these issues is essential for improving both performance assessment and clinical and preventive intervention design.

In this Special Issue, the contributions encompass studies on upper- and lower-limb asymmetry and their functional consequences, such as the relationship between morphological asymmetry and grip strength [8] and the association between temporal–spatial gait asymmetry and dynamic balance in children with hemiplegic cerebral palsy [9]. Several papers explore symmetry in sports and physical activities, including bilateral landing mechanics under different flooring conditions [10], laterality and spatial asymmetry in elite football penalty kicks [11], and the relationship between handedness and scoring success in handball [12]. Additional contributions address applied and methodological aspects, such as asymmetries in school furniture alignment and anthropometry [13], the influence of core muscle endurance on frontal-plane movement symmetry during indoor walking or cycling [14], and the dependence of asymmetry outcomes on the test used and player category in male volleyball players [15].

Taken together, these studies provide new information regarding (i) how symmetry and asymmetry manifest in human movement, (ii) how they can be reliably assessed, and (iii) how they may inform training strategies, injury prevention, and rehabilitation programs. We hope that this Special Issue will serve as a useful reference for researchers, clinicians, and practitioners interested in advancing the understanding and application of symmetry-related concepts in motor control, sports, and rehabilitation.

Author Contributions: Conceptualization, A.d.S.F. and F.V.d.A.; writing—original draft preparation, A.d.S.F. and F.V.d.A.; writing—review and editing, A.d.S.F. and F.V.d.A.; supervision, A.d.S.F. and

F.V.d.A.; project administration, A.d.S.F. and F.V.d.A.; funding acquisition, A.d.S.F. and F.V.d.A. All authors have read and agreed to the published version of the manuscript.

Funding: This research was funded by the Fundação Carlos Chagas Filho de Amparo à Pesquisa do Estado do Rio de Janeiro, under grant numbers E-26/204.369/2024; the Conselho Nacional de Desenvolvimento Cientifico e Tecnológico, under grant numbers 408494/2023-9 and 311827/2025-0; and the Coordenação de Aperfeiçoamento de Pessoal de Nível Superior, under grant numbers 88881.708719/2022-01 and 88887.708718/2022-00.

Data Availability Statement: No new data were created.

Conflicts of Interest: The authors declare no conflicts of interest.

References

1. Fatima Zaidi, Z. Body Asymmetries: Incidence, Etiology and Clinical Implications. *Aust. J. Basic Appl. Sci.* **2011**, *5*, 2157–2191.
2. Bullock-Saxton, J. Postural Alignment in Standing: A Repeatability Study. *Aust. J. Physiother.* **1993**, *39*, 25–29. [CrossRef] [PubMed]
3. Siegler, A.; Pick, C.G.; Been, E. Differences in Body Positional Bilateral Symmetry between Stance and Supine Positions, and the Impact of Attention and Awareness on Postural Symmetry. *Gait Posture* **2019**, *68*, 476–482. [CrossRef] [PubMed]
4. Prado-Rico, J.M.; Duarte, M. Asymmetry of Body Weight Distribution During Quiet and Relaxed Standing Tasks. *Motor. Control* **2019**, *23*, 553–565. [CrossRef] [PubMed]
5. Kutilek, P.; Svoboda, Z.; Viteckova, S.; Hana, K.; Vana, Z. Evaluation of the Effect of Heel Lift on Postural Stability and Symmetry of Muscle Activity. *J. Back Musculoskelet. Rehabil.* **2017**, *30*, 1037–1044. [CrossRef] [PubMed]
6. Haddad, M. Motor Asymmetry in Football: Implications for Muscular Power, Balance, and Injury Prevention. *Symmetry* **2024**, *16*, 1485. [CrossRef]
7. Ojeda-Aravena, A.; Warnier-Medina, A.; Brand, C.; Morales-Zúñiga, J.; Orellana-Lepe, G.; Zapata-Bastias, J.; Tuesta, M. Relationship between Body Composition Asymmetry and Specific Performance in Taekwondo Athletes: A Cross-Sectional Study. *Symmetry* **2023**, *15*, 2087. [CrossRef]
8. Alecu, S.; Onea, G.A.; Badau, D.; Badau, A.; Nechita, F. Influence of Bilateral Upper Limb Morphological Asymmetry on Grip Strength Related to Gender in Non-Athlete University Students. *Symmetry* **2026**, *18*, 122. [CrossRef]
9. Elnaggar, R.K. The Magnitude of Temporal–Spatial Gait Asymmetry Is Related to the Proficiency of Dynamic Balance Control in Children with Hemiplegic Cerebral Palsy: An Analytical Inquiry. *Symmetry* **2024**, *16*, 1274. [CrossRef]
10. Busuttil, N.A.; Dunn, M.; Roberts, A.H.; Viglione, A.J.; Middleton, K.J. Effects of Sports Flooring on Peak Ground Reaction Forces During Bilateral Drop Landings. *Symmetry* **2025**, *17*, 2045. [CrossRef]
11. Cidre-Fuentes, P.; González-Harcevnicow, M.A.; Prieto-Lage, I. Laterality, Shot Direction and Spatial Asymmetry in Decisive Penalty Kicks: Evidence from Elite Men's Football. *Symmetry* **2025**, *17*, 1570. [CrossRef]
12. Laxdal, A.; Ivarsson, A.; Thorgeirsson, S.; Haugen, T. Born to Score? The Relationship between Left-Handedness and Success from the 7-Meter Line. *Symmetry* **2022**, *14*, 2163. [CrossRef]
13. Paramés-González, A.; Gutiérrez-Santiago, A.; Fernandes, A.; Braga, A.C.; Costa, N.; Carneiro, P.; Reguera-López-de-la-Osa, X.; Prieto-Lage, I. Asymmetry in the Alignment of School Furniture and Anthropometric Measures: A Comparative Study Between Two Schools in Spain and Portugal. *Symmetry* **2024**, *16*, 1588. [CrossRef]
14. Park, J.-H.; Kim, A.-R.; Kim, S.-H.; Kim, K.B.; Park, K.-N. Association of Core Muscle Endurance with Weekly Workout Time, Speed, and the Symmetry of Frontal Core Motion during Indoor Walking and Cycling. *Symmetry* **2022**, *14*, 2333. [CrossRef]
15. Iglesias-Caamaño, M.; Álvarez-Yates, T.; Carballo-López, J.; Cuba-Dorado, A.; García-García, Ó. Is Asymmetry Different Depending on How It Is Calculated? *Symmetry* **2022**, *14*, 2195. [CrossRef]

https://doi.org/10.3390/sym18010166

symmetry

MDPI

Article

Influence of Bilateral Upper Limb Morphological Asymmetry on Grip Strength Related to Gender in Non-Athlete University Students

Stefan Alecu [1], Gheorghe Adrian Onea [1,*], Dana Badau [2], Adela Badau [1] and Florentina Nechita [1]

[1] Department of Physical Education and Special Motricity, Transilvania University of Brașov, 500024 Brașov, Romania; alecu.stefan@unitbv.ro (S.A.); adela.badau@unitbv.ro (A.B.); florentina.nechita@unitbv.ro (F.N.)

[2] Department of Motor Performance, Transilvania University of Brașov, 500024 Brașov, Romania; dana.badau@unitbv.ro

* Correspondence: onea.gheorghe@unitbv.ro

Abstract

Bilateral morphological asymmetry of the upper limbs may influence grip strength even in semi-active young adults. Understanding this relationship is important for identifying early neuromuscular imbalances with implications for ergonomics and rehabilitation. This study aimed to examine associations between upper limb anthropometric characteristics and grip strength in non-athlete students, considering gender and manual dominance. The sample included 192 healthy university students (110 females, 82 males; mean age 19.92 ± 1.4 years) without prior sports training. Thirteen bilateral anthropometric parameters of the upper limbs were assessed, including hand and palm dimensions, segmental lengths, and arm and forearm circumferences, along with grip strength measured by dynamometry in two positions: arm extended and arm flexed at 90°. Statistical analysis revealed significant differences in forearm length, arm and forearm circumferences, and grip strength ($p < 0.001$). The dominant limb consistently demonstrated higher grip strength, with mean differences of approximately 2 kg. Male participants showed higher absolute values for all morphological and functional variables, whereas stronger correlations between distal upper-limb morphology and grip strength were observed in females. These findings indicate that, despite largely symmetric skeletal dimensions, moderate functional asymmetries exist and grip strength is influenced primarily by local muscular development rather than overall limb size.

Keywords: bilateral symmetry; upper limb morphology; grip strength; dynamometry; anthropometric analysis; functional asymmetry

1. Introduction

Upper limb asymmetry is a phenomenon frequently observed even among apparently healthy young adults, such as students. These differences are influenced by factors such as manual dominance [1–3], daily habits of preferential hand use, and individual neuromuscular development. In students, who spend much time in sedentary or repetitive academic activities (writing, keyboard use), these asymmetries can become more pronounced, especially in the absence of functional balance generated through physical activity [4,5].

Over time, even minor differences can affect posture, movement ergonomics, or the distribution of muscular effort [6,7]. The evaluation of these discrepancies becomes essential not only for understanding general functional status, but also for preventing chronic imbalances, especially in the context of transition to active professional life. In this sense, analyzing morphological and functional symmetry offers a valuable perspective on body balance [8–10] and can support personalized interventions in physical education, rehabilitation, or even vocational selection.

This asymmetry can be morphological in nature: differences in length, width, circumference between the right and left limb; or functional: differences in grip strength [11,12], mobility, or motor coordination capacity.

Functional and morphological asymmetries [13–15] of the upper limbs [16,17] are frequently encountered among students, even in the absence of an active sports regimen. Differences in strength, length, or muscle mass differences between the right and left limbs can be explained both by manual dominance (right-handed vs. left-handed) and by disproportionate daily use of the dominant hand. These asymmetries can influence functional performance [18–20], posture, and risk of injury in daily life.

Studies have shown that even in the absence of sports practice, manual dominance [21] plays a significant role in the development of functional asymmetries [22], favoring the strength and muscle mass of the dominant limb [23]. Although the young adult population, especially students, is often perceived as functionally homogeneous, recent research suggests significant asymmetries in the upper limbs [24,25], even in the absence of sports training. These asymmetries can be explained by lateral dominance, but also by repetitive daily activities that involve preference for using a single hand.

A significant correlation was observed between grip strength and body segment size among young adults, with notable bilateral variations [26]. Additionally, Burdukiewicz et al. (2020) reported significant differences between athletes and sedentary students regarding the symmetry of grip strength and muscle mass [23]. This evidence suggests that asymmetry is a common characteristic among the university population and can serve as a useful indicator for functional assessments or corrective interventions. Moreover, differences in arm circumference, palm length, or grip strength [27] reflect not only natural anatomical variations, but also individual functional adaptations. Evaluating these asymmetries among non-athlete students [28–30] offers a valuable perspective on musculoskeletal balance status, with implications for prevention of imbalances, performance optimization, and early rehabilitation interventions. In this context, the present study aims to examine the relationship between bilateral morphological symmetry [31,32] and functional symmetry, offering a complete analysis of potential links between anthropometric parameters and force generation capacity at the upper limb level [33,34].

Based on an analysis of the specialized literature, we identified several gaps in research on morphological asymmetry and manual grip strength. Research on the influence of bilateral morphological asymmetry of the upper limbs on grip strength among non-athlete students has significant gaps that limit the validity and applicability of the results. From a methodological perspective, the current literature reflects a lack of standardization in measurement protocols, manifested through variability in criteria for defining significant morphological asymmetry and absence of consensus on relevant anthropometric parameters. Methods for evaluating grip strength vary in testing positions, number of attempts, and recovery intervals, compromising study comparability. Additionally, inadequate control of confounding variables such as lateral dominance, impact of sedentary behaviors, and occupational factors specific to the university population is observed. Aspects related to gender differences in the manifestation of asymmetry and the implications of

https://doi.org/10.3390/sym18010122

different equipment and technology use on asymmetric development are insufficiently explored. From a conceptual perspective, uncertainties persist regarding cause-and-effect relationships between morphological asymmetry and strength differences, and underlying neuromotor mechanisms remain unclear [35,36]. The lack of longitudinal studies prevents understanding of the dynamics of asymmetry development during the university period and the long-term implications for musculoskeletal health. The practical applicability of results is limited by the absence of reference values specific to the population and correlations with functional performance in daily activities [37,38]. These limitations highlight the necessity for standardized protocols, longitudinal studies, and a clear definition of the significance of morphological asymmetry in the context of the non-athletic university population. Research that integrates in a unified manner segmental measurements (lengths, circumferences, spans) with bilateral dynamometric evaluations among non-athlete students is relatively scarce and focuses especially on the assessment of motor components.

The present study makes an original contribution by investigating the relationship between morphological and functional symmetry of the upper limbs in a population of young adult students who are semi-active or moderately active and do not engage in competitive sports. Unlike the existing literature, which predominantly focuses on elite athletes or clinical populations with musculoskeletal pathologies, this research adopts an integrated approach, combining detailed segmental anthropometric assessments with bilateral dynamometric measurements performed in two distinct positions. The main innovative aspects consist of: characterization of asymmetries in an intermediate population, representative of the majority of young people in academic settings, thus providing reference values for healthy individuals with moderate physical activity; simultaneous and correlated analysis of morphological parameters (segmental measurements of arm, forearm, and hand) and functional parameters (bilateral grip strength), enabling the identification of direct structure–function relationships; dynamometric evaluation in two different biomechanical contexts, which allows the detection of position-dependent asymmetries and provides a more comprehensive picture of bilateral functional capacity; and comparative analysis between sexes, both for morphological and functional asymmetries, as well as for the correlations between them, aspects insufficiently investigated in previous studies on non-athletic populations. This multidimensional approach allows not only the identification of the degree of morphological and functional asymmetry at the upper limb level, but also testing the hypothesis that structural asymmetries correlate with functional ones and that these relationships may differ according to gender and the biomechanical context of the assessment.

The study aims to identify bilateral differences and correlations between segmental anthropometric dimensions of the upper limbs and grip strength (evaluated by dynamometry) in young adults, students, and non-athletes, by gender and manual dominance.

The study hypothesis was that asymmetry at the upper-limb level influences manual grip strength (dynamometry) in non-athlete students and that these effects differ by gender and manual prevalence.

2. Materials and Methods

2.1. Participants

The study included 192 participants, of which 110 were female and 82 were male, with ages ranging between 18 and 26 years (mean age: 19.92 years). The mean body mass and height of the total sample were 66.7 ± 13.8 kg and 170.9 ± 8.9 cm, respectively. In the female group, mean body mass was 60.1 ± 10.0 kg and mean height was 165.7 ± 6.6 cm, while in the male group, mean body mass was 75.5 ± 13.4 kg and mean height was 177.9 ± 6.2 cm. All subjects were volunteers recruited from students at a higher education institution,

 https://doi.org/10.3390/sym18010122

clinically healthy and fit for physical effort, without musculoskeletal conditions or other contraindications that could influence anthropometric or functional measurements. Clinically healthy students who did not have musculoskeletal conditions or other medical contraindications were included in the study and provided informed consent for participation. An additional inclusion criterion was the subjects' declaration that they do not regularly practice physical exercise or are not registered with any sports club, to ensure the sample was representative of the non-athlete young adult population.

2.2. Study Design

This study took place from 27 March to 15 May 2025, with the objective of measuring anthropometric parameters and dynamometry of the subjects' upper limbs. The anthropometric measurement sessions were conducted under similar conditions and with the same measurement instruments for all participants. The order of performing anthropometric measurements was identical for all subjects. Anthropometric measurements of the upper limbs were performed bilaterally (both on the right side and on the left side). All measurements were performed in the faculty's physical education halls. Participants wore specific sports clothing and were instructed not to engage in intense physical exercise for at least 12 h before testing. Each measurement was performed with the same apparatus and equipment by the same evaluator to reduce inter-rater variability. Each parameter was recorded once, except for dynamometry, where the highest value from two attempts was taken into consideration. Data were centralized in digital format (Microsoft Excel 2021), processed, and statistically validated before final analysis. For measurements, the following were used: a stadiometer for height, a digital scale, a digital caliper, a flexible measuring tape, and a digital dynamometer.

The study was approved by the Ethical Board of the Faculty of Physical Education and Mountain Sports of Transilvania University of Brasov under the document no. 101/26.03.2025.

2.3. Measurements

For this study, 15 measurements were performed: 13 anthropometric and two motor. All anthropometric measurements were conducted according to standardized anthropometric procedures based on internationally accepted guidelines. Anatomical landmarks, participant positioning, and measurement techniques were defined in advance and applied consistently across all participants, following principles commonly used in anthropometric research and comparable to those described in the International Society for the Advancement of Kinanthropometry (ISAK) framework and were performed with standardized equipment as follows: for anthropometric evaluations of widths and lengths at the hand and palm levels, we used a Mitutoyo 500-196-30 digital caliper with a measurement range of 0–100 mm, a reading precision of 0.01 mm, and a tolerance of ± 0.01 mm. Segmental lengths (such as forearm length, arm length, and upper limb length) were measured with a flexible tape measure (150 cm), and circumferences were measured with an ergonomic tape measure featuring a self-tightening mechanism to ensure uniform application without tissue compression (Seca 201, Seca GmbH, Hamburg, Germany). For evaluations of hand grip strength, a GRIPX digital dynamometer was used, with a capacity of 90 kg, electronic. Although no formal validation study specific to this model has been published, digital hand dynamometers have been shown to provide reliable and valid measurements of grip strength in young adult populations when standardized testing protocols are applied. Similar digital dynamometry devices [39] have been widely used in previous studies investigating grip strength and its relationship with anthropometric parameters in students and healthy adults, supporting the methodological adequacy of this approach.

 https://doi.org/10.3390/sym18010122

Anthropometric evaluations:

— Height—distance between the vertex (top of the head) and the sole level (support surface) in orthostatic position.
— Palm width—direct distance from the most lateral point of the second metacarpal head to the most medial point of the fifth metacarpal head.
— Palm width (thumb included)—direct distance from the most lateral point of the first metacarpal head to the most medial point of the fifth metacarpal head.
— Palm length—distance between the styloid line and the proximal phalanges between the middle and ring fingers.
— Hand length—distance between the styloid line and dactylion.
— Palm span—distance between the proximal phalanges of the little finger and the distal phalanges of the thumb, with fingers extended at maximum angles.
— Upper limb length—distance between the acromion and dactylion in orthostatic position, with the upper limb completely extended.
— Forearm length—distance between the lateral epicondyle of the humerus (olecranon) and the styloid process of the radius (styloid line where the prominence at the wrist is felt).
— Wrist circumference—on the styloid line, at the level of the styloid processes of the radius and ulna bones, i.e., at the two lateral bony prominences of the wrist. The measuring tape is placed exactly over these points, applied firmly but without compressing tissue, completely surrounding the joint.
— Relaxed forearm circumference—with the subject in orthostatic position and palm oriented anteriorly, half the distance between the lateral epicondyle and styloid process is determined, and the measuring tape is wrapped around the forearm.
— Flexed forearm circumference—with the subject in orthostatic position and palm oriented anteriorly, the subject is instructed to clench the fist and tense the forearm muscles in isometry, with the measuring tape wrapped around the forearm.
— Relaxed arm circumference—with the subject in orthostatic position and arm bent at the elbow at 90°, half the distance between the acromion and olecranon is determined and the measuring tape is wrapped at mid-distance, around the arm, without compressing tissues.
— Flexed arm circumference—with the subject in orthostatic position and arm bent at the elbow at 90°, in isometry, the subject is instructed to simultaneously clench the fist and flex the biceps, with the measuring tape wrapped around the arm.

Arm and forearm circumferences were analyzed as absolute anthropometric measures. Corrected circumferences adjusted for adipose tissue were not calculated, as skinfold thickness were not included in the study protocol. We aim to evaluate practical, field-based anthropometric indicators commonly used in educational and ergonomic settings, rather than estimates of isolated muscle cross-sectional area.

Motor evaluations:

— Dynamometry with extended arm—with the subject in orthostatic position with extended arm and palm oriented anteriorly, the subject is instructed to squeeze the dynamometer as hard as possible, without compensatory movements. Two attempts are performed with a 30 s pause between them. The highest value is recorded, expressed in kilograms (kg).
— Dynamometry with arm bent at 90°—with the subject in orthostatic position, with arm bent at the elbow joint at 90°, the subject is instructed to squeeze the dynamometer as hard as possible, without compensatory movements. Two attempts are performed

https://doi.org/10.3390/sym18010122

with a 30 s pause between them. The highest value is recorded, expressed in kilograms (kg).

2.4. Statistical Analysis

Analyses were performed in IBM SPSS Statistics for Windows Version 26; IBM Corp., Armonk, NY, USA, with significance threshold $\alpha = 0.05$ (bilateral). The statistical power of the sample was calculated with G*Power 3.1.9.4. a priori/post hoc, indicating that for the main functional effects ($d^z \approx 0.46$–0.53), a sample of ~28–37 participants would ensure power ≥ 0.80 ($\alpha = 0.05$), while for smaller morphological effects, ~102 participants would be necessary. Our study included $n = 192$ subjects; the achieved power for key results was ≈ 1.00 (functional) and ~0.97 (e.g., forearm length), confirming sample adequacy for detecting effects of interest. For each bilateral variable, descriptive indicators (mean, standard deviation, minimum, maximum, skewness, kurtosis, CV%) and 95% confidence intervals (95% CI) were calculated. Left–right differences were evaluated using paired Student's *t*-tests; comparisons between genders were evaluated using independent Student's *t*-tests; and bivariate Pearson correlations were calculated between manual anthropometric measurements and dynamometric strength measurements. The coefficient of variation (CV) was calculated to highlight group homogeneity. We also calculated Cohen's d effect size. For paired tests, effect size was expressed through the statistical parameter Cohen's (d^z). The Pearson correlation coefficient was used to identify correlations between anthropometric indicators and grip strength; analyses were performed separately by gender (M/F), laterality (left/right), and posture (extension/flexion at 90°), with 12 anthropometric indicators per set. For identifying differences between the left and right upper segments. Before performing inferential analyses, the assumptions for parametric testing were examined. Data normality was assessed using the Shapiro–Wilk test, and homogeneity of variance was evaluated using Levene's test. These analyses supported the use of parametric statistical procedures.

3. Results

In Table 1, the analysis of the fourteen bilaterally measured anthropometric parameters reveals consistent differences between genders, especially in linear dimensions and circumferences, as well as good body symmetry between the left and right sides. Regarding palm width, the mean values were 9.45 ± 0.61 cm in the female group and 10.63 ± 1.09 cm in the male group, with the difference confirmed by 95% confidence intervals. Hand length showed a similar trend, with means of 17.18 ± 0.94 cm in the female group and 18.89 ± 1.04 cm in the male group. For circumferences, values were significantly higher in the male group. At the wrist level, the mean was 15.6 cm in the female group and 17.4 cm in the male group. The difference was accentuated at the flexed arm, where the female group had a mean circumference of 27.9 ± 3.95 cm, while the male group had a mean circumference of 32.9 ± 4.24 cm.

Table 1. Descriptive statistics of the 14 parameters according to laterality and gender.

Parameters	Hand	Gender	Min.	Max.	X	SD	CI 95%		Kurtosis	CV%
							Lower	Upper		
Palm width with thumb (cm)	L	F	8.00	10.70	9.444	0.613	9.329	9.559	−0.509	6.496
		M	8.80	12.00	10.615	0.773	10.447	10.782	−0.564	7.283
	R	F	8.00	10.80	9.436	0.607	9.322	9.550	−0.276	6.438
		M	8.80	12.00	10.549	0.729	10.391	10.707	−0.219	6.912

https://doi.org/10.3390/sym18010122

Table 1. *Cont.*

Parameters	Hand	Gender	Min.	Max.	X	SD	CI 95%		Kurtosis	CV%
							Lower	Upper		
Palm width (cm)	L	F	7.00	9.00	7.869	0.456	7.783	7.954	−0.294	5.800
		M	6.50	10.00	8.743	0.592	8.615	8.871	1.644	6.773
	R	F	6.70	9.50	7.790	0.493	7.697	7.882	0.709	6.325
		M	7.30	10.00	8.650	0.548	8.531	8.769	−0.354	6.338
Palm length (cm)	L	F	8.50	11.30	9.768	0.615	9.653	9.883	−0.300	6.292
		M	8.50	12.00	10.556	0.675	10.410	10.702	0.751	6.394
	R	F	8.40	11.30	9.720	0.602	9.607	9.833	0.025	6.195
		M	8.80	12.00	10.513	0.682	10.366	10.661	0.495	6.489
Hand length (cm)	L	F	15.50	19.80	17.406	0.881	17.240	17.571	−0.244	5.060
		M	17.00	21.00	18.744	0.908	18.547	18.940	−0.637	4.846
	R	F	13.20	20.50	17.380	1.052	17.182	17.577	1.720	6.051
		M	16.80	20.50	18.635	0.928	18.435	18.836	−0.664	4.979
Hand span (cm)	L	F	16.50	22.50	18.986	1.306	18.741	19.231	−0.386	6.881
		M	18.20	24.60	21.240	1.415	20.934	21.547	−0.437	6.662
	R	F	13.30	22.00	18.882	1.452	18.609	19.154	0.989	7.692
		M	18.80	25.30	21.223	1.463	20.906	21.540	−0.223	6.894
Forearm length (cm)	L	F	20.00	28.00	24.112	1.604	23.811	24.413	−0.251	6.651
		M	22.70	28.50	25.941	1.299	25.660	26.223	−0.389	5.007
	R	F	20.00	29.00	24.369	1.801	24.031	24.707	−0.027	7.392
		M	22.50	29.50	26.328	1.518	26.000	26.657	0.064	5.765
Arm length (cm)	L	F	27.00	36.00	31.490	1.946	31.125	31.855	−0.369	6.180
		M	27.00	43.50	33.155	3.161	32.471	33.839	1.185	9.534
	R	F	26.00	36.50	31.453	2.021	31.074	31.833	−0.220	6.427
		M	26.00	43.00	32.939	3.463	32.189	33.689	0.374	10.514
Wrist circumference (cm)	L	F	13.40	19.00	15.649	1.293	15.406	15.891	−0.428	8.263
		M	15.50	19.00	17.068	0.936	16.866	17.271	−0.806	5.487
	R	F	13.40	19.50	15.645	1.280	15.405	15.885	0.004	8.179
		M	15.50	19.00	17.082	0.889	16.889	17.274	−0.738	5.207
Relaxed forearm circumference (cm)	L	F	17.80	30.00	23.507	2.312	23.073	23.941	0.594	9.835
		M	22.00	32.50	26.937	2.299	26.439	27.434	−0.192	8.534
	R	F	18.40	30.30	23.876	2.266	23.451	24.302	0.454	9.491
		M	22.40	33.50	27.279	2.336	26.774	27.785	0.027	8.564
Tense forearm circumference (cm)	L	F	18.70	31.00	24.099	2.276	23.672	24.526	0.687	9.445
		M	23.00	33.50	27.677	2.407	27.156	28.198	−0.057	8.696
	R	F	19.10	31.00	24.476	2.269	24.050	24.902	0.451	9.270
		M	23.20	34.50	28.059	2.368	27.546	28.571	0.191	8.438
Relaxed arm circumference (cm)	L	F	18.40	38.30	26.106	3.644	25.422	26.790	1.093	13.957
		M	21.50	42.00	29.882	3.894	29.039	30.724	0.367	13.030
	R	F	19.20	38.00	26.330	3.664	25.642	27.018	1.226	13.916
		M	21.50	43.00	29.987	3.767	29.171	30.802	1.368	12.563

https://doi.org/10.3390/sym18010122

Table 1. *Cont.*

Parameters	Hand	Gender	Min.	Max.	X	SD	CI 95%		Kurtosis	CV%
							Lower	Upper		
Tense arm circumference	L	F	21.00	41.00	27.760	3.702	27.065	28.455	1.492	13.335
		M	23.70	47.00	32.868	4.239	31.951	33.786	1.158	12.898
	R	F	21.60	41.00	28.008	3.626	27.328	28.689	1.284	12.946
		M	24.70	47.00	33.118	4.306	32.186	34.050	1.029	13.002
Extended arm dynamometry (kg)	L	F	13.50	36.80	23.865	5.019	22.923	24.807	−0.346	21.031
		M	20.00	64.50	36.450	10.606	34.154	38.746	−0.211	29.097
	R	F	12.50	44.50	25.862	6.232	24.692	27.032	−0.237	24.096
		M	19.90	69.30	38.728	11.707	36.194	41.262	−0.315	30.230
Bend arm dynamometry (kg)	L	F	9.90	39.90	23.098	5.515	22.063	24.134	0.201	23.877
		M	18.50	71.00	35.235	11.295	32.791	37.680	1.150	32.055
	R	F	10.70	36.80	25.004	6.059	23.866	26.141	−0.536	24.233
		M	19.50	62.90	37.505	11.122	35.097	39.912	−0.615	29.656

Min—minimum; Max—maximum; X—mean; SD—standard deviation; CI 95%—confidence interval 95%; CV—coefficient of variation; L—left; R—right; F—female; M—male.

Comparison of the left and right sides did not reveal significant differences. Means were almost identical across genders, and confidence intervals overlapped. For example, forearm length was 23.9 ± 1.46 cm (left) and 24.3 ± 1.54 cm (right) in the female group, and 26.3 ± 1.69 cm and 26.6 ± 1.75 cm in the male group, respectively. These results suggest stable structural symmetry between the two sides of the body.

The most evident differences between genders were observed in strength tests (dynamometry). In the extended arm test, the female group had a mean of 23.7 ± 7.6 kgf, while the male group had a mean of 36.1 ± 9.1 kgf. In the bent arm test, results were similar: 22.9 ± 8.2 kgf in the female group and 36.2 ± 8.5 kgf in the male group. Maximum values reached 55 kgf in the female group and over 70 kgf in the male group, illustrating not only the difference between means, but also greater variability in the male group.

The study results confirm that the male group presents significantly larger dimensions and strength than the female group, while bilateral symmetry is well preserved. These observations are relevant to both the sports and ergonomic domains, as well as to clinical applications, where assessment of body proportions and muscle strength can contribute to understanding functional performance and injury risk.

Descriptive characteristics of the sample showed clear differences between genders. Male participants presented higher mean body mass and height compared to female participants, while age distribution was comparable between groups.

The comparative analysis in Table 2 of upper limb anthropometric parameters revealed clear differences between genders, with consistently higher values in the male group across all linear dimensions and circumferences, as well as in muscle strength tests. At the same time, no notable differences were observed between the left and right sides, suggesting stable bilateral body symmetry. Regarding palm and hand dimensions, the male group showed superior values compared to the female group across width, length, and span. For example, hand length was approximately 2 cm greater in the male group, and palm span exceeded female group values by about 2.5 cm, with differences statistically confirmed ($p < 0.001$). At the level of proximal segments (forearm and arm), differences became even more evident. Mean forearm length was greater in the male group by approximately 2 cm, and arm length

https://doi.org/10.3390/sym18010122

by 4–5 cm. These results confirm that gender differences are expressed more markedly at the level of large bone and muscle structures. The most evident contrasts were observed in circumferences. Under conditions of muscle contraction, arm circumference in the male group exceeded that of the female group by nearly 5 cm, reflecting differences in muscle mass.

Table 2. Paired sample test analysis of the anthropometrics and dynamometric parameters according to handedness.

Parameters	Hand	ΔX	SD	95% CI		t	*p*	d
				LL	UL			
Palm width with thumb (cm)	L	9.945	0.895	9.818	10.073	1.392	0.166	0.100
	R	9.911	0.859	9.789	10.034			
Palm width (cm)	L	8.240	0.675	8.144	8.336	3.537	0.001	0.255
	R	8.156	0.669	8.060	8.251			
Palm length (cm)	L	10.098	0.756	9.990	10.206	1.857	0.065	0.134
	R	10.055	0.749	9.948	10.162			
Hand length (cm)	L	17.971	1.114	17.813	18.130	1.412	0.160	0.102
	R	17.913	1.176	17.745	18.080			
Hand span (cm)	L	19.942	1.757	19.692	20.192	1.049	0.296	0.076
	R	19.882	1.856	19.618	20.147			
Forearm length (cm)	L	24.880	1.746	24.631	25.128	−3.697	0.000	−0.267
	R	25.198	1.943	24.922	25.475			
Arm length (cm)	L	32.204	2.656	31.825	32.582	1.116	0.266	0.081
	R	32.091	2.817	31.690	32.492			
Wrist circumference (cm)	L	16.261	1.346	16.069	16.453	−0.357	0.721	−0.026
	R	16.270	1.334	16.080	16.460			
Relaxed forearm circumference (cm)	L	24.995	2.862	24.588	25.403	−9.482	0.000	−0.684
	R	25.355	2.848	24.949	25.760			
Tense forearm circumference (cm)	L	25.650	2.926	25.233	26.067	−10.007	0.000	−0.722
	R	26.030	2.911	25.615	26.444			
Relaxed arm circumference (cm)	L	27.749	4.186	27.154	28.345	−1.996	0.047	−0.144
	R	27.922	4.120	27.335	28.508			
Tense arm circumference (cm)	L	30.005	4.721	29.333	30.677	−2.506	0.013	−0.181
	R	30.258	4.722	29.586	30.931			
Straight arm dynamometry (kg)	L	29.232	10.050	27.801	30.662	−6.345	0.000	−0.458
	R	31.360	10.989	29.796	32.925			
Bended arm dynamometry (kg)	L	28.274	10.373	26.798	29.751	−7.315	0.000	−0.527
	R	30.346	10.565	28.842	31.850			

CI—interval of confidence; LL—lower limit; UL—upper limit; SD—standard deviation; d—effect size; L—left; R—right; t—Student's t; *p*—significance level; d = Cohen's effect size, statistical significance was set at $p < 0.01$.

In muscle strength tests, the contrast was clearly superior in favor of the male group. Dynamometry values showed differences of approximately 12–13 kg between genders in both the extended-arm and bent-arm tests ($p < 0.001$). Thus, the data support the idea of a pronounced gender difference at the upper-limb level, as evidenced by higher values in the male group across all anthropometric parameters and muscle strength. However, bilateral

https://doi.org/10.3390/sym18010122

symmetry between the left and right sides suggests that these differences are uniformly distributed and do not depend on laterality.

The results of the paired test in Table 3, analyzed separately for the female group and male group, show that bilateral asymmetry in anthropometric parameters is reduced for most measurements. For the forearm, both in the female and male groups, consistent differences were evident between the left and right sides. In the female group, right forearm circumference was approximately 0.33 cm greater at rest and 0.35 cm in contraction, both differences being highly significant ($p < 0.001$). The effect was of moderate magnitude (d ≈ −0.66 and −0.72), suggesting clear muscular development of the dominant side. In the male group, values were similar: +0.40 cm at rest and +0.42 cm in contraction on the right side ($p < 0.001$, d ≈ −0.63 and −0.74). The largest differences appeared in grip strength. The female group had on average 2.1 kg more strength in the right hand, both with extended arm and bent arm. Both results were extremely statistically significant ($p < 0.001$) and with moderate effects (d ≈ −0.5). In the male group, differences were almost identical: −2.14 kg for extended arm and −2.09 kg for bent arm, confirming the same tendency of superior strength on the right side ($p < 0.001$).

Table 3. Paired Sample test analysis between left and right hands of the anthropometrics and dynamometric parameters according to gender.

Parameters	Gender	ΔX	SD	95% CI		t	*p*	d
				LL	UL			
Palm width with thumb (cm)	F	−0.010	0.286	−0.044	0.064	−0.367	0.714	0.035
	M	−0.066	0.395	−0.021	0.152	−1.509	0.135	−0.167
Palm width (cm)	F	−0.079	0.335	0.015	0.141	−2.452	0.016	−0.236
	M	0.927	0.327	0.020	0.164	−2.564	0.012	2.835
Palm length (cm)	F	−0.043	0.273	−0.009	0.094	1.641	0.104	−0.158
	M	0.042	0.372	−0.039	0.124	1.037	0.303	0.113
Hand length (cm)	F	0.022	0.700	−0.110	0.154	0.327	0.744	0.031
	M	0.109	0.350	0.031	0.185	2.808	0.006	0.311
Hand span (cm)	F	0.091	0.824	−0.065	0.246	1.156	0.250	0.110
	M	0.017	0.730	−0.143	0.177	0.212	0.833	0.023
Forearm length (cm)	F	−0.268	1.072	−0.471	0.065	−2.624	0.010	−0.250
	M	−0.387	1.345	−0.682	−0.090	−2.601	0.011	−0.288
Arm length (cm)	F	0.037	1.081	−0.168	0.240	0.353	0.725	0.034
	M	0.216	1.746	−0.167	0.599	1.119	0.266	0.124
Wrist circumference (cm)	F	−0.005	0.337	−0.069	0.058	−0.169	0.866	−0.015
	M	−0.014	0.352	−0.090	0.064	−0.344	0.731	−0.040
Relaxed forearm circumference (cm)	F	−0.371	0.477	−0.462	−0.281	−8.165	0.000	−0.778
	M	−0.342	0.585	−0.471	−0.214	−5.299	0.000	−0.585
Tense forearm circumference (cm)	F	−0.378	0.480	−0.469	−0.287	−8.254	0.000	−0.788
	M	−0.382	0.584	−0.510	−0.253	−5.919	0.000	−0.654
Relaxed arm circumference (cm)	F	−0.222	0.855	−0.384	−0.061	−2.732	0.007	−0.260
	M	−0.105	1.544	−0.444	0.234	−0.615	0.540	−0.068
Tense arm circumference (cm)	F	−0.255	0.745	−0.396	−0.114	−3.592	0.000	−0.342
	M	−0.250	1.967	−0.682	0.182	−1.151	0.253	−0.127

https://doi.org/10.3390/sym18010122

Table 3. *Cont.*

Parameters	Gender	ΔX	SD	95% CI		t	*p*	d
				LL	UL			
Straight arm dynamometry (kg)	F	−2.017	3.769	−2.729	−1.305	−5.613	0.000	−0.535
	M	−2.278	5.638	−3.517	−1.039	−3.660	0.000	−0.404
Bended arm dynamometry (kg)	F	−1.923	3.147	−2.518	−1.328	−6.409	0.000	−0.611
	M	−2.270	4.784	−3.320	−1.218	−4.295	0.000	−0.475

CI—interval of confidence; LL—lower limit; UL—upper limit; SD—standard deviation; M—male; F—female; d—effect size; t—Student's t; *p*—significance level; statistical significance was set at $p < 0.01$.

Analysis of effect size (Cohen's d) revealed important differences between structural and functional parameters of the upper limbs. For most linear measurements (palm length and width, hand or arm length, wrist circumference), d values were very small (below 0.2), indicating small effects and little practical relevance. This result suggests bilateral structural symmetry, with minimal differences between the left and right sides. In contrast, for muscular and functional parameters, effects of moderate to large magnitude were identified. Forearm circumference showed effects ranging from d = −0.63 to d = −0.74, in both the female and male groups, in the relaxed state and during contraction. These values indicate consistent differences in favor of the right side, reflecting more pronounced muscular development in the dominant forearm. Additionally, grip strength presented moderate effects (Cohen's d $\approx$ −0.5 to −0.6), confirming a clear functional difference between the literalities. The approximately 2 kg difference in favor of the right hand, observed in both the female and male groups, demonstrates the practical relevance of these effects. Thus, Cohen's d values suggest that while bone and articular dimensions remain largely symmetric, bilateral asymmetry with practical effect manifests especially at the muscular and functional level, being closely linked to manual dominance.

Mean grip strength values for dominant and non-dominant upper limbs in female and male participants, measured with the arm extended and with the arm flexed at 90°. Dominant limbs correspond to the right side for the majority of participants.

Table 4, Figure 1, reveals distinct association patterns, with relative structural symmetry and asymmetric, gender-dependent functionality. Among female participants, Pearson correlation analysis reveals significant associations between forearm and wrist circumference variables and grip strength. For example, flexed forearm circumference (right) correlates strongly with grip strength in extended position (r = 0.262, $p = 0.006$) and bent position (r = 0.305, $p = 0.001$). In contrast, wrist and relaxed forearm circumference show significant bilateral correlations, with r values ranging between 0.256 and 0.350 ($p < 0.01$). These results suggest that distal segmental muscular development is an efficient predictor of functional strength in the female population. Notably, palm width also correlates significantly with right-sided grip strength (r = 0.436, $p < 0.001$), suggesting a possible structural influence of the support base on force transmission.

In male subjects, relationships are less pronounced. Still, significant correlations are observed between relaxed forearm circumference (left) and grip strength in both positions (r $\approx$ 0.29, $p < 0.01$), as well as between flexed arm circumference and grip strength (r = 0.264, $p = 0.016$). This functional asymmetry in the expression of correlations suggests a different biomechanical model between genders. In the female group, grip strength appears more dependent on distal parameters (forearm, wrist). In contrast, in the male group, strength is distributed more diffusely and influenced by additional neuromuscular factors not directly reflected in segmental morphology. Additionally, the consistency of significant

 https://doi.org/10.3390/sym18010122

relationships on the right side in both the male group and female group confirms functional lateral dominance, with associated morphological adaptations.

Table 4. Pearson correlations between dynamometric measurements and anthropometric parameters in relation to gender and handedness.

Parameters	Gender	Side	Straight Arm Dynamometry		Bended Arm Dynamometry	
			r	*p*	r	*p*
Palm width with thumb (cm)	M	L	−0.125	0.262	−0.014	0.902
		R	0.025	0.821	0.092	0.409
	F	L	0.144	0.134	0.157	0.104
		R	0.281 **	0.003	0.256 **	0.007
Palm width (without thumb) (cm)	M	L	0.121	0.279	0.133	0.235
		R	0.130	0.245	0.143	0.200
	F	L	0.274 **	0.004	0.327 **	0.001
		R	0.437 **	0.000	0.422 **	0.000
Palm length (cm)	M	L	0.198	0.076	0.192	0.085
		R	0.210	0.064	0.205	0.071
	F	L	0.175	0.069	0.153	0.113
		R	0.205 *	0.033	0.152	0.115
Hand length (cm)	M	L	0.112	0.317	0.104	0.350
		R	0.033	0.766	0.027	0.810
	F	L	0.223 *	0.020	0.193 *	0.044
		R	0.275 **	0.004	0.270 **	0.005
Hand span (cm)	M	L	0.181	0.104	0.150	0.180
		R	0.191	0.087	0.161	0.147
	F	L	0.093	0.337	0.162	0.092
		R	0.068	0.482	0.104	0.281
Forearm length (cm)	M	L	0.236 *	0.032	0.211	0.063
		R	0.253 *	0.021	0.225 *	0.049
	F	L	0.212 *	0.027	0.218 *	0.023
		R	0.085	0.378	0.100	0.303
Arm length (cm)	M	L	0.298 **	0.006	0.266 *	0.015
		R	0.303 **	0.005	0.270 *	0.014
	F	L	0.008	0.930	0.007	0.944
		R	0.106	0.272	0.065	0.501
Wrist circumference (cm)	M	L	0.195	0.080	0.176	0.113
		R	0.210	0.064	0.201	0.070
	F	L	0.326 **	0.001	0.354 **	0.000
		R	0.265 **	0.005	0.326 **	0.001

https://doi.org/10.3390/sym18010122

Table 4. *Cont.*

Parameters	Gender	Side	Straight Arm Dynamometry		Bended Arm Dynamometry	
			r	*p*	r	*p*
Relaxed forearm circumference (cm)	M	L	0.317 **	0.004	0.309 **	0.005
		R	0.328 **	0.003	0.315 **	0.004
	F	L	0.319 **	0.001	0.285 **	0.003
		R	0.259 **	0.007	0.307 **	0.001
Tense forearm circumference (cm)	M	L	0.333 **	0.002	0.322 **	0.003
		R	0.342 **	0.002	0.328 **	0.003
	F	L	0.342 **	0.000	0.275 **	0.004
		R	0.265 **	0.005	0.308 **	0.001
Relaxed arm circumference (cm)	M	L	0.355 **	0.001	0.349 **	0.001
		R	0.365 **	0.001	0.358 **	0.001
	F	L	0.265 **	0.005	0.238 *	0.013
		R	0.157	0.103	0.260 **	0.006
Tense arm circumference (cm)	M	L	0.384 **	0.000	0.378 **	0.000
		R	0.395 **	0.000	0.389 **	0.000
	F	L	0.254 *	0.008	0.228 *	0.017
		R	0.174	0.070	0.251 *	0.008

* M—group of male; F—group of female; L—left; R—right; r = Pearson correlation coefficient. ** Statistical significance was set at $p < 0.01$. ** Correlation magnitude was interpreted as follows: trivial (r < 0.10), small (r = 0.10–0.29), moderate (r = 0.30–0.49), and large (r ≥ 0.50).

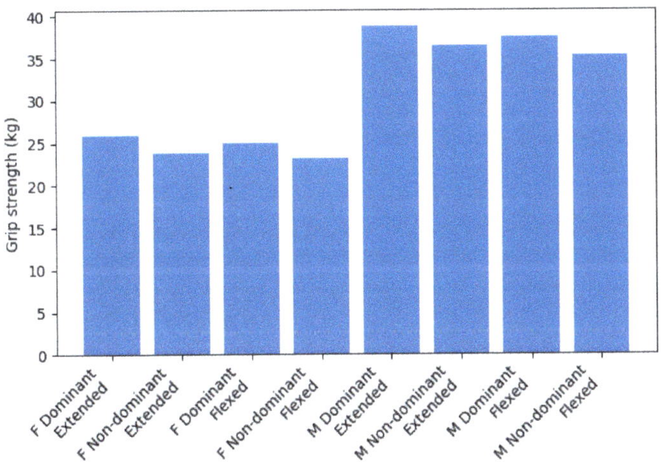

Figure 1. Grip strength differences between dominant and non-dominant upper limbs by sex and testing position.

4. Discussion

The main purpose of this study was to investigate the relationship between segmental morphological dimensions of the upper limbs and grip strength in the context of bilateral asymmetry in young non-athletes. By analyzing differences in the upper limbs and between

https://doi.org/10.3390/sym18010122

genders, the study aimed to highlight possible functional or structural imbalances. The obtained results contribute to understanding how the preferential use of one upper limb and anthropometric characteristics influence functional performance, even in the absence of organized sports training.

Statistical analysis revealed significant differences between the dominant and non-dominant limbs in functional parameters, especially grip strength, which showed clear superiority of the dominant limb across testing positions. This functional asymmetry appears as a systematic and recurrent phenomenon, reflecting neuromuscular adaptations induced by predominant unilateral use. At the morphological level, although most parameters showed relatively well-preserved symmetry, significant discrepancies were observed in forearm and arm length and circumference, suggesting localized structural adaptation, most likely due to daily activity. Gender differences were also evident: the male group showed higher values across all segmental dimensions and grip strength, whereas in the female group, stronger correlations were observed between morphological characteristics and functional performance.

Interestingly, although subjects in the male group had significantly greater strength, subjects in the female group demonstrated more consistent and significant correlations between segmental dimensions (forearm circumference, palm width) and grip strength. This aspect suggests a more predictable functional model in the female group, where morphological development is more closely linked to functional performance. In contrast, in the male group, strength appears influenced by additional factors, possibly neurophysiological or hormonal, that are not directly reflected in anthropometric dimensions.

The study results validate the research hypothesis regarding the link between morphological characteristics and functional performance, offering an essential scientific foundation for using anthropometric evaluations in predicting grip strength. Therefore, Table 4 not only supports the inclusion of gender and laterality differences in predictive models of neuromuscular function but also highlights an essential functional asymmetry that must be considered in ergonomics, sports medicine, and rehabilitation.

The results of this study offer a detailed perspective on bilateral asymmetries [40,41] of the upper limbs among university students not involved in organized sports activity. Although specialized literature tends to emphasize symmetry as a functional and morphological ideal in healthy populations [42,43], our data indicate statistically significant discrepancies, especially in functional parameters (grip strength) and, to a lesser extent, in some morphological parameters, such as forearm and arm circumferences.

The most prominent form of asymmetry observed was functional, reflected in grip strength differences between upper limbs [34,44,45]. This observation is in accordance with specialized literature, which emphasizes that manual dominance leads to preferential use of one limb, inducing functional hypertrophy and, consequently, greater strength in the dominant limb [1,46–48]. Thus, differences of approximately 2 kg on average in dynamometry tests between left and right limbs support the hypothesis of systematic and constant functional asymmetry among non-athlete subjects. It is noteworthy that these differences are not significantly influenced by the biomechanical testing position (extended arm vs. bent arm), suggesting that the determining factors are neuromuscular rather than strictly biomechanical.

In contrast to functional parameters, most morphological parameters did not show significant bilateral differences. Notable exceptions were forearm length and forearm and arm circumferences (both relaxed and flexed), where differences of up to 0.5–0.8 cm were observed between limbs. These discrepancies, although relatively small in absolute terms,

https://doi.org/10.3390/sym18010122

are statistically significant and indicate differentiated muscular adaptation in the dominant limb, likely associated with repetitive, sustained use in daily activities.

These results are consistent with those reported by [23], who observed a similar pattern of lateralized functional hypertrophy in athletes and the general population. However, unlike studies focused on performance athletes, where morphological differences are much more pronounced [49,50], the population studied in this case presents relatively well-preserved structural symmetry, indicating a moderate influence of daily behavior on segmental morphology.

Gender analysis revealed clear differentiation regarding both morphological and functional parameters. The male group recorded significantly superior values for all strength indicators and for most segmental dimensions (forearm, arm, palm). This result aligns with recent meta-analyses that confirm physiological and hormonal gender differences in muscle mass and strength performance [26,34].

A central aspect emerging from current literature is the influence of asymmetric use on neuromuscular and morphological development of the upper limbs. Even in untrained populations, such as those investigated in this study, manual dominance determines differential stimulation of the musculature, with measurable effects on local strength and tone. It has been demonstrated that laterality significantly influences grip performance and that this difference directly correlates with segmental circumferences, especially at the forearm level, in healthy young people. This finding supports our results regarding the localized character of functional adaptations [26].

Additionally, the analysis of sex differences in the present study is in accordance with the literature, suggesting a differentiated distribution of muscle mass and gender-specific neuromuscular recruitment. Studies also emphasize that female and male group use different strategies during muscular effort, even within upper-limb muscle groups [51]. Thus, our results, which indicate a more predictable relationship between morphology and strength in the female group, can be explained by more linear neuromuscular control and a more uniform distribution of muscle mass involved in gripping actions. In contrast, in the male group, factors such as neural impulse, joint stiffness, or anabolic hormones may play a more important role in force generation, independent of morphology.

Beyond the explanatory dimension, these findings also have applicative implications in the field of ergonomics and rehabilitation. Grip strength, as an easily evaluated parameter, can serve as an early indicator of neuromuscular imbalances or risk of dysfunctions associated with excessive unilateral use. A specialized study highlights the need to assess strength by position and by professional activity, showing that position differences can have a reduced impact in the presence of marked functional dominance, as our data indicate [52]. We consider that dynamometric and segmental anthropometric evaluations can constitute a valuable tool for evaluating correlations between anthropometric and motor parameters, as predictive elements of body symmetries and harmonious physical development in young people [53–57].

4.1. Practical Implications of the Study

The identification and quantification of these functional and morphological asymmetries has major practical importance. In the context of physical education and the prevention of postural imbalances, these results can guide personalized interventions to balance functional loads on the upper limbs. Additionally, in ergonomics and occupational medicine, these data can serve as a basis for adapting repetitive tasks (e.g., typing, lifting objects, desk work) to be evenly distributed between limbs, thereby reducing the risk of unilateral overload and musculoskeletal dysfunctions. At the same time, in neuromus-

 https://doi.org/10.3390/sym18010122

cular rehabilitation and sports medicine, evaluating these asymmetries can inform the development of differentiated training or recovery protocols that account for laterality, gender-related morphological variation, and segmental muscular adaptations.

4.2. Limitations and Future Directions

The study has several limitations that must be considered. First, the investigated population consisted exclusively of young, healthy students, limiting the generalizability of the results to other age groups or clinical populations. Second, electromyographic or imaging measurements were not included to highlight differences in muscle activation or tissue composition. In the future, integrating these methods could provide a more detailed picture of the mechanisms underlying the identified functional asymmetries. Additionally, the direct influence of daily activities or a history of physical exercise on symmetry was not analyzed, an aspect that could be explored in future longitudinal research. Evaluation of other body segments is also necessary to understand whether the asymmetry pattern is localized or generalized.

5. Conclusions

The research results partially confirmed the initial hypothesis: general morphological asymmetry is reduced and does not significantly influence functional parameters, but specific muscle segments and local circumferences clearly correlate with grip strength. Paired analyses showed constant functional asymmetry in favor of the right side, where grip strength was approximately 2 kg greater, both in the female group and male group. In parallel, forearm circumference (at rest and in contraction) presented significant differences between left and right, with moderate effects. Pearson correlations revealed positive links between forearm and arm circumference and grip strength, especially in the female group. Regarding gender differences, the male group recorded higher values for all segmental dimensions and grip strength, confirming evident functional and morphological differences between genders. However, in the female group, correlations between morphological and functional parameters were more consistent, indicating a closer relationship between structure and performance.

The study confirms that bilateral structural asymmetry (lengths and widths) does not significantly affect functional performance, whereas muscular asymmetry, especially at the forearm level, directly influences grip strength. These findings validate the research hypothesis and support the proposed title, highlighting that the influence of morphology on strength is localized and functional rather than global. The results can contribute to better understanding of the relationship between morphology and function in young semi-active populations. They can support preventive or rehabilitation programs oriented toward functional balancing of the upper limbs. Additionally, the study contributes to filling the scientific gap regarding the relationship between the bilateral morphology of the upper limbs and their functionality among the semi-active student population. These results can form the basis for preventive or corrective interventions that address functional and structural imbalances from the early stages of adult life.

Author Contributions: Conceptualization, S.A., G.A.O., D.B., A.B. and F.N.; methodology, S.A., G.A.O. and D.B.; validation, S.A., G.A.O. and D.B.; formal analysis, S.A., G.A.O. and D.B.; investigation, S.A., G.A.O. and D.B.; data curation, S.A., G.A.O. and D.B.; writing—original draft preparation, S.A., G.A.O., D.B., A.B. and F.N.; writing—review and editing, S.A., G.A.O., D.B., A.B. and F.N. All authors have read and agreed to the published version of the manuscript.

Funding: This research received no external funding.

 https://doi.org/10.3390/sym18010122

Institutional Review Board Statement: The study was conducted in accordance with the Declaration of Helsinki, and the protocol was approved by the Ethical Board of the Faculty of Physical Education and Mountain Sports of Transilvania University of Brasov under the document no.101/26.03.2025 on 26 March 2025.

Informed Consent Statement: Informed consent was obtained from all subjects involved in the study.

Data Availability Statement: The original contributions are included in the article.

Conflicts of Interest: The authors declare no conflicts of interest.

References

1. Purves, D.; White, L.E.; Andrews, T.J. Manual asymmetry and handedness. *Proc. Natl. Acad. Sci. USA* **1994**, *91*, 5030–5032. [CrossRef]
2. Papadatou-Pastou, M.; Ntolka, E.; Schmitz, J.; Martin, M.; Munafò, M.R.; Ocklenburg, S.; Paracchini, S. Human handedness: A meta-analysis. *Psychol. Bull.* **2020**, *146*, 481–524. [CrossRef] [PubMed]
3. Geserick, G.; Krocker, K.; Schmeling, T. On manual laterality (handedness) in humans and its forensic significance—A literature review. *Arch. Kriminol.* **2016**, *239*, 145–166. [PubMed]
4. Biddle, S.J.; Asare, M. Physical activity and mental health in children and adolescents: A review of reviews. *Br. J. Sports Med.* **2011**, *45*, 886–895. [CrossRef]
5. de Greeff, J.W.; Bosker, R.J.; Oosterlaan, J.; Visscher, C.; Hartman, E. Effects of physical activity on executive functions, attention and academic performance in preadolescent children: A meta-analysis. *J. Sci. Med. Sport* **2018**, *21*, 501–507. [CrossRef] [PubMed]
6. Dos Anjos, F.V.; Ghislieri, M.; Cerone, G.L.; Pinto, T.P.; Gazzoni, M. Changes in the distribution of muscle activity when using a passive trunk exoskeleton depend on the type of working task: A high-density surface EMG study. *J. Biomech.* **2022**, *130*, 110846. [CrossRef] [PubMed]
7. De Las Casas, H.; Bianco, S.; Richter, H. Targeted muscle effort distribution with exercise robots: Trajectory and resistance effects. *Med. Eng. Phys.* **2021**, *94*, 70–79. [CrossRef]
8. Rdzanek, J.; Świątek, J.; Wychowański, M. Correlation Between Foot Arch Parameters and Body Balance Performance. *Ortop. Traumatol. Rehabil.* **2022**, *24*, 263–272. [CrossRef] [PubMed]
9. Almurdi, M.M. Postural balance and functional mobility in relation to BMI and body composition among female students at a College of Applied Medical Sciences: A cross-sectional study. *Clinics* **2024**, *79*, 100401. [CrossRef]
10. Cui, X.; Fu, B.; Liu, S.; Cheng, Y.; Wang, X.; Zhao, T. Study on the Difference of Human Body Balance Stability Regulation Characteristics by Time-Frequency and Time-Domain Data Processing Methods. *Int. J. Environ. Res. Public Health* **2022**, *19*, 14078. [CrossRef]
11. Pavlovic, R.; Vrcic, M. Hand Grip Strength in Students: Differences in the Gender Dimorphism. *Int. J. Phys. Educ. Fit. Sports* **2021**, *10*, 13–21. [CrossRef]
12. Sukanta, S. Evaluation of Anthropometric Traits, Somatotype and Grip Strength of College Students. *Online Int. Interdiscip. Res. J.* **2013**, *3*, 94–103.
13. Bisiacchi, P.; Cainelli, E. Structural and functional brain asymmetries in the early phases of life: A scoping review. *Brain Struct. Funct.* **2022**, *227*, 479–496. [CrossRef]
14. Pirttiniemi, P. Normal and increased functional asymmetries in the craniofacial area. *Acta Odontol. Scand.* **1998**, *56*, 342–345. [CrossRef]
15. Specht, K.; Wigglesworth, P. The functional and structural asymmetries of the superior temporal sulcus. *Scand. J. Psychol.* **2018**, *59*, 74–82. [CrossRef]
16. Goble, D.J.; Lewis, C.A.; Brown, S.H. Upper limb asymmetries in the utilization of proprioceptive feedback. *Exp. Brain Res.* **2006**, *168*, 307–311. [CrossRef]
17. Adamo, D.E.; Scotland, S.; Martin, B.J. Upper limb kinesthetic asymmetries: Gender and handedness effects. *Neurosci. Lett.* **2012**, *516*, 188–192. [CrossRef] [PubMed]
18. Hasan, S. Effects of plyometric vs. strength training on strength, sprint, and functional performance in soccer players: A randomized controlled trial. *Sci. Rep.* **2023**, *13*, 4256. [CrossRef] [PubMed]
19. Guadagnin, E.C.; Barbieri, F.A.; Simieli, L.; Carpes, F.P. Is Muscular And Functional Performance Related To Gait Symmetry In Older Adults? A Systematic Review. *Arch. Gerontol. Geriatr.* **2019**, *84*, 103899. [CrossRef] [PubMed]
20. Charest, J.; Grandner, M.A. Sleep and Athletic Performance: Impacts on Physical Performance, Mental Performance, Injury Risk and Recovery, and Mental Health: An Update. *Sleep Med. Clin.* **2022**, *17*, 263–282. [CrossRef] [PubMed]

https://doi.org/10.3390/sym18010122

21. Badau, A.; Badau, D. Identifying the Differences in Symmetry of the Anthropometric Parameters of the Upper Limbs in Relation to Manual Laterality between Athletes Who Practice Sports with and without a Ball. *Symmetry* **2024**, *16*, 558. [CrossRef]
22. Lucki, N.C.; Nicolay, C.W. Phenotypic plasticity and functional asymmetry in response to grip forces exerted by intercollegiate tennis players. *Am. J. Hum. Biol.* **2007**, *19*, 566–577. [CrossRef] [PubMed]
23. Burdukiewicz, A.; Pietraszewska, J.; Andrzejewska, J.; Chromik, K.; Stachoń, A. Asymmetry of Musculature and Hand Grip Strength in Bodybuilders and Martial Artists. *Int. J. Environ. Res. Public Health* **2020**, *17*, 4695. [CrossRef] [PubMed]
24. Trojano, L. Hemispheric asymmetries in the control of upper limb movements. *Handb. Clin. Neurol.* **2025**, *208*, 393–405. [CrossRef]
25. Velarde-Sotres, Á.; Bores-Cerezal, A.; Mecías-Calvo, M.; Barcala-Furelos, M.; Aparicio-Obregón, S.; Calleja-González, J. Detection of Upper Limb Asymmetries in Athletes According to the Stage of the Season—A Longitudinal Study. *Int. J. Environ. Res. Public Health* **2022**, *19*, 849. [CrossRef] [PubMed]
26. Zaccagni, L.; Toselli, S.; Bramanti, B.; Gualdi-Russo, E.; Mongillo, J.; Rinaldo, N. Handgrip Strength in Young Adults: Association with Anthropometric Variables and Laterality. *Int. J. Environ. Res. Public Health* **2020**, *17*, 4273. [CrossRef]
27. Leyk, D.; Gorges, W.; Ridder, D.; Wunderlich, M.; Rüther, T.; Sievert, A.; Essfeld, D. Hand-grip strength of young men, women and highly trained female athletes. *Eur. J. Appl. Physiol.* **2007**, *99*, 415–421. [CrossRef] [PubMed]
28. Edwards, B.; Froehle, A. Examining the incidence of reporting mental health diagnosis between college student athletes and non-athlete students and the impact on academic performance. *J. Am. Coll. Health* **2023**, *71*, 69–75. [CrossRef] [PubMed]
29. Campa, F.; Coratella, G. Athlete or Non-Athlete? This Is the Question in Body Composition. *Front. Physiol.* **2021**, *12*, 814572. [CrossRef] [PubMed]
30. Clemente, F.M.; Nikolaidis, P.T.; Martins, F.M.; Mendes, R.S. Weekly physical activity patterns of university students: Are athletes more active than non-athletes? *Springerplus* **2016**, *5*, 1808. [CrossRef]
31. Savriama, Y.; Klingenberg, C.P. Beyond bilateral symmetry: Geometric morphometric methods for any type of symmetry. *BMC Evol. Biol.* **2011**, *11*, 280. [CrossRef]
32. Kleipool, R.P.; Vuurberg, G.; Stufkens, S.A.S.; van der Merwe, A.E.; Oostra, R.J. Bilateral symmetry of the subtalar joint facets and the relationship between the morphology and osteoarthritic changes. *Clin. Anat.* **2020**, *33*, 997–1006. [CrossRef] [PubMed]
33. Melo, F.C.M.; de Lima, K.K.F.; Silveira, A.P.K.F.; de Azevedo, K.P.M.; Dos Santos, I.K.; de Medeiros, H.J.; Leitão, J.C.; Knackfuss, M.I. Physical Training and Upper-Limb Strength of People with Paraplegia: A Systematic Review. *J. Sport Rehabil.* **2019**, *28*, 288–293. [CrossRef] [PubMed]
34. Nuzzo, J.L.; Pinto, M.D. Sex Differences in Upper- and Lower-Limb Muscle Strength in Children and Adolescents: A Meta-Analysis. *Eur. J. Sport Sci.* **2025**, *25*, e12282. [CrossRef]
35. Smith, B.A.; Lang, C.E. Sensor Measures of Symmetry Quantify Upper Limb Movement in the Natural Environment Across the Lifespan. *Arch. Phys. Med. Rehabil.* **2019**, *100*, 1176–1183. [CrossRef]
36. Chan, M.S.; Sigward, S. Prioritizing limb loading improves symmetry during dual-tasking in individuals following anterior cruciate ligament reconstruction. *Front. Sports Act. Living* **2023**, *5*, 1090694. [CrossRef] [PubMed]
37. Missaoui, B.; Portero, P.; Bendaya, S.; Hanktie, O.; Thoumie, P. Posture and equilibrium in orthopedic and rheumatologic diseases. *Neurophysiol. Clin.* **2008**, *38*, 447–457. [CrossRef] [PubMed]
38. Fanelli, A.; Laddomada, T.; Sacchelli, M.; Allegri, M. Acute and chronic pain management in sport medicine: An expert opinion looking at an alternative mechanism-based approach to the pharmacological treatment. *Minerva Anestesiol.* **2023**, *89*, 468–477. [CrossRef] [PubMed]
39. Benton, M.J.; Spicher, J.M.; Silva-Smith, A.L. Validity and reliability of handgrip dynamometry in older adults: A comparison of two widely used dynamometers. *PLoS ONE* **2022**, *17*, e0270132. [CrossRef]
40. Hartley, C.; Taylor, N.; Chidley, J.; Baláš, J.; Giles, D. Handedness, Bilateral, and Interdigit Strength Asymmetries in Male Climbers. *Int. J. Sports Physiol. Perform.* **2023**, *18*, 1390–1397. [CrossRef]
41. López-Fernández, J.; García-Unanue, J.; Sánchez-Sánchez, J.; Colino, E.; Hernando, E.; Gallardo, L. Bilateral Asymmetries Assessment in Elite and Sub-Elite Male Futsal Players. *Int. J. Environ. Res. Public Health* **2020**, *17*, 3169. [CrossRef]
42. Lopez, C. Against Symmetry Fundamentalism. *Erkenntnis* **2025**, *90*, 1821–1845. [CrossRef]
43. Buetow, S.; Wallis, K. The Beauty in Perfect Imperfection. *J. Med. Humanit.* **2019**, *40*, 389–394. [CrossRef] [PubMed]
44. Martinez, R.; Assila, N.; Goubault, E.; Begon, M. Sex differences in upper limb musculoskeletal biomechanics during a lifting task. *Appl. Ergon.* **2020**, *86*, 103106. [CrossRef]
45. Guirelli, A.R.; Dos Santos, J.M.; Cabral, E.M.G.; Pinto, J.P.C.; De Lima, G.A.; Felicio, L.R. Relationship between upper limb physical performance tests and muscle strength of scapular, shoulder and spine stabilizers: A cross-sectional study. *J. Bodyw. Mov. Ther.* **2021**, *27*, 612–619. [CrossRef] [PubMed]
46. Shenoy, P.; Varadhan, S.K.M. Task demands modulate distal limb handedness: A comparative analysis of prehensile synergies of the dominant and non-dominant hand. *Sci. Rep.* **2024**, *14*, 25565. [CrossRef] [PubMed]

 https://doi.org/10.3390/sym18010122

47. Loprinzi, P.D.; Franklin, J.; Farris, A.; Ryu, S. Handedness, Grip Strength, and Memory Function: Considerations by Biological Sex. *Medicina* **2019**, *55*, 444. [CrossRef]
48. Badau, D.; Steff, N.; Badau, A.; Stoica, M. Enhancing Lower-Limb Simple and Choice Reaction Time and Spatial Orientation in Junior Basketball Players by Implementing Fitlight Technology in Sports Training. *Percept. Mot. Ski.* **2025**, *132*, 1392–1417. [CrossRef] [PubMed]
49. Ramos, S.; Volossovitch, A.; Ferreira, A.P.; Barrigas, C.; Fragoso, I.; Massuça, L. Differences in Maturity, Morphological, and Fitness Attributes Between the Better- and Lower-Ranked Male and Female U-14 Portuguese Elite Regional Basketball Teams. *J. Strength Cond. Res.* **2020**, *34*, 878–887. [CrossRef] [PubMed]
50. Gonzalez, M.P.; Montalvo, S.; Dietze-Hermosa, M.; Gomez, M.; Dorgo, S. A Comparison of Morphological, Jump, and Sprint Kinematic Asymmetries in Division I Track and Field Athletes. *Int. J. Exerc. Sci.* **2023**, *16*, 1306–1319. [CrossRef]
51. Mehta, R.K.; Rhee, J. Revealing Sex Differences during Upper and Lower Extremity Neuromuscular Fatigue in Older Adults through a Neuroergonomics Approach. *Front. Neuroergon.* **2021**, *2*, 663368. [CrossRef] [PubMed]
52. Panoutsakopoulos, V.; Schwarz, A.V.; Merkou, E.; Savvoulidis, S.; Kotzamanidou, M.C.; Papadakis, Z. Handgrip Strength Asymmetry in Female Basketball Players: The Critical Role of Arm Position and the Challenge of Statistical Power. *Sports* **2025**, *13*, 279. [CrossRef] [PubMed]
53. Dumitrescu, R.; Stoica, A.; Aducovschi, D.; Ganciu, M.; Sakizlian, R.; Moise, D.; Gozu, B.; Sakizlian, M.; Mihaiu, C.; Leştaru, M. Study about the place of sports activities in students free time budget. *Eur. Proc. Soc. Behav. Sci.* **2017**, *23*, 1605–1613. [CrossRef]
54. Moraru, L.; Mirea, O.; Toader, D.; Berceanu, M.; Soldea, S.; Munteanu, A.; Donoiu, I.; Raicea, V. Lower Limit of Normality of Segmental Multilayer Longitudinal Strain in Healthy Adult Subjects. *J. Cardiovasc. Dev. Dis.* **2024**, *11*, 102. [CrossRef]
55. Martoma, A.M.; Tache, S.; Aronescu Cârjan, P.A.; Kiss, M.; Vadan, A. Flavonoizii şi capacitatea de efort fizic. *Palestrica Mileniului III* **2010**, *11*, 346–349.
56. Cavlak, U.; Jurak, I.; Rusu, L.; Aytutuldu, G.K.; Telebuh, M.; Budeanca-Babolea, O.B.; Kısa, E.P.; Bertic, Z.; Ilie, E.N.; Grozdek Covcic, G.; et al. An Analysis of Factors Associated with Chronic Musculoskeletal Pain, Pain Management Preferences, Coping Strategies, and Health-Related Quality of Life Among Older People: A Cross-Cultural Study. *J. Pain Res.* **2025**, *18*, 4915–4934. [CrossRef] [PubMed]
57. Rusu, L.; Cosma, G.; Calina, M.L.; Dragomir, M.M.; Marin, M. Evaluation of two muscle training programs by assessment of the muscle tone. *Sci. Sports* **2015**, *30*, e65–e72. [CrossRef]

 https://doi.org/10.3390/sym18010122

symmetry

MDPI

Article

Effects of Sports Flooring on Peak Ground Reaction Forces During Bilateral Drop Landings

Nicholas A. Busuttil [1,2,3,*], Marcus Dunn [4], Alexandra H. Roberts [2], Anthony J. Viglione [3] and Kane J. Middleton [2]

[1] First Year College, Victoria University, Melbourne 8001, Australia
[2] Sport, Performance and Nutrition Research Group, School of Allied Health, Human Services and Sport, La Trobe University, Melbourne 3086, Australia; k.middleton@latrobe.edu.au (K.J.M.)
[3] Performance Science, Research and Innovation, The Movement Institute, Melbourne 3042, Australia
[4] School of Energy, Geoscience, Infrastructure and Society, Heriot-Watt University, Edinburgh EH14 4AS, UK
* Correspondence: nick.busuttil@vu.edu.au; Tel.: +61-3-9919-5147

Abstract: With continued advancement in flooring technology, modular sports flooring tiles have emerged as a potential alternative flooring solution for sports performance. However, there is limited empirical evidence regarding their effects on ground reaction forces in landing tasks (GRFs). Therefore, the aim of this study was to assess the effects of flooring surface on peak GRFs during bilateral drop landings. Eighteen physically active adults (10 males, 8 females) completed three bilateral drop landings from a 50 cm height across each of three flooring types: modular sport tiles, athletic track, and bare force plates, measuring contacts from both the left and right limb. GRFs were captured using two in-ground force platforms, normalised to body mass, and then analysed using a linear mixed-effects model with post-hoc comparisons where significant interactions were recorded. Peak anterior GRFs were significantly lower in the modular tiles compared with athletic sports track and bare metal surfaces ($p < 0.001$, $\eta^2_p \geq 0.430$). Additionally, anterior ($p = 0.048$, $\eta^2_p = 0.040$), lateral ($p < 0.001$, $\eta^2_p = 0.280$), and vertical ($p = 0.001$, $\eta^2_p = 0.100$) GRFs were significantly greater in the right limb compared with the left limb within each flooring surface condition. Future research should investigate sport-specific movement patterns and long-term adaptations associated with training on modular surfaces to assess their potential role in enhancing performance and mitigating injuries.

Keywords: kinetics; ground reaction forces; drop landings; surface; flooring

1. Introduction

Playing surface is a critical factor that can influence athletic performance and injury risk across various sports. Contemporary sports surfaces are designed to incorporate natural, synthetic, or a combination of materials to enhance overall performance experience and mitigate injury risk during athletic movements [1]. Over recent decades, artificial turf has been increasingly adopted as an alternative to natural grass, particularly in field-based sports, due to its durability and consistent playing conditions [1]. However, despite its widespread implementation, there is limited evidence on the effectiveness of artificial surfaces in replicating natural surface characteristics regarding injury prevalence and severity [2,3].

Since the initial adoption of artificial playing surfaces in field sports competitions (e.g., soccer), other sports have subsequently adopted similar innovations. Indoor sports such as basketball utilise maple hardwood as the primary playing surface due to its stability for dribbling, reduced likelihood to expand and contract with changes in temperature, and shock absorption properties [4]. Biomechanical analyses have demonstrated that key performance metrics, including peak vertical ground reaction force, power output, and jump height, remain comparable across surface types during athletic tasks such as countermovement and depth jumps [5]. In contrast, one study that assessed consecutive ankle jumps (no active use of the knees) and standard countermovement jumps on varied thicknesses of athletic track reported a significant main effect of flooring surface, whereby the vertical instantaneous loading rate ($N{\cdot}s^{-1}{\cdot}Kg^{-1}$) during landing was reduced in three of four conditions (flooring thickness; SF1: 0.014 m; SF2: 0.007 m; SF3: 0.011 m; SF4: 0.018 m) compared with the reference condition (SF0) [6]). Contact time between jumps was also reported to be significantly longer from SF2 to SF1 [6]. It is evident that sports surfaces are a rapidly evolving and well-researched area for athletic performance; however, a deeper understanding of the biomechanical implications of sports surface engagement is crucial.

Modern sport surfaces continually evolve with the aim to improve the floorings' force absorption ability, shoe–surface friction characteristics, vertical deformation, and energy transfer to an individual's body [7,8]. For example, landing tasks are a primary movement in basketball (single- or double-legged landing) as a result of dribbling or attempting to score from open play. High surface friction, coupled with rapid decelerating movements, have been suggested to increase lower limb injury rate by two-fold [9]. One new and emerging technology is modular flooring tiles, which are an interlocking thermoplastic material joining system that links sport floor tiles together with the goal of reducing vertical and horizontal braking forces exerted on a tile by an athlete [10], a key component of anterior cruciate ligament (ACL) ruptures, an injury that has a high incidence rate across high school, collegiate, and professional athletes (59%) [11]. Anecdotally, modular sports flooring tiles have been an increasingly utilised surface for sporting activities and can be used as an alternate playing surface. It is suggested that using modular flooring surfaces may result in lowered peak ground reaction forces (GRFs), and loading rates of modular tiles are aligned to reduce the risk of lower-body injuries [12]. Using this flooring technology may enable a gradual load exposure during injury rehabilitation, suggesting reduced ground contact landing forces compared with traditional surfaces [12]. However, the effect of modular flooring surfaces on peak ground reaction forces compared with traditional and commonly used flooring surfaces during landing tasks is unclear.

Therefore, the aim of this study was to quantify the effects of different flooring surfaces (modular sport tiles, athletic track, and bare force plates) on peak ground reaction forces during bilateral drop landings. It was hypothesised that the athletic track surface would exhibit lower peak vertical ground reaction forces compared with the modular sporting tile surface and bare (metal force plate) flooring.

2. Materials and Methods

2.1. Participants

Eighteen healthy male and female adults (mean $\pm$ SD; 10 males, 8 females; age: 24.6 $\pm$ 2.0 years, height: 1.74 $\pm$ 0.1 m, mass: 72.5 $\pm$ 7.5 kg) participated in the study. All participants were injury-free for at least six months at the time of testing and had previous experience in performing general athletic tasks (jumping and landing). Written informed consent was captured before participation in the study, which was approved by the La Trobe University Institutional Human Research Ethics Committee (#HEC21082).

2.2. Testing Procedure and Protocol

All testing was completed in a Sports Biomechanics Laboratory (La Trobe University, Melbourne, Australia). Participants were provided a demonstration of the experimental design and were afforded time to complete their individual warm-up.

Participants were required to perform bilateral drop landings (DL) from a 50 cm height on three surfaces. The surfaces included two bare force plates as a reference condition (SF1), indoor Mondo athletic track (Sportflex Super X 720™ K39, Mondo, Alba, Italy; SF2) affixed to the two force plates, and two interlocked modular sports flooring tiles per plate (MSF Elite PRO, MSF Sports, Melbourne, Australia; SF3) mounted to the two force plates. All drop landings were performed on top of the force plates. The participants were required to perform three DLs for each flooring condition. Between each trial, participants were afforded 30 s of rest and 3 min of rest between conditions. Participants were instructed to stand still until prompted to "step out" with their right leg from the starting position, then instructed to land as if they were dropping down from a height. All DLs were performed with their hands on their hips, and participants were prompted to land naturally and hold the landing for two seconds. Drop landings were performed with each foot striking an individual force plate (mounted 13.5 mm below the laboratory floor), where GRFs were captured with two ground-embedded force platforms (BMS6001200; Advanced Mechanical Technology, Inc., Watertown, MA, USA; 1000 Hz). To achieve consistent drop heights, an offset height of 13.5 mm was applied to the SF2 and SF3 conditions as the athletic flooring track (custom-fitted according to the force plate dimensions) and modular flooring tiles (two interlocked per force plate) were positioned on top of the in-ground force plate, which fitted flush within the grooves of the custom force plate dimensions cutout and laboratory floor level (Figure 1). If complete foot contacts on individual landing surfaces were not achieved during the DL trial, the participants were required to repeat the landing until a total of three successful DLs across each surface condition were collected. Surface type (SF1, SF2, and SF3) was counterbalanced during data collection. Kinetic data were captured using Vicon Nexus software (Vicon Motion Systems Ltd., Oxford, UK, V 2.16.1). Between each surface condition, the force plates were zeroed. Raw force data were not filtered to avoid producing errors in peak GRFs similar to previous research [13]. Ground reaction forces, normalised to body mass ($N \cdot Kg^{-1}$), were analysed using the conventional coordinate system, where force in the anterior direction was defined as positive along the x-axis (Fx), lateral to the right as positive along the y-axis (Fy), and vertical force upward as positive along the z-axis (Fz).

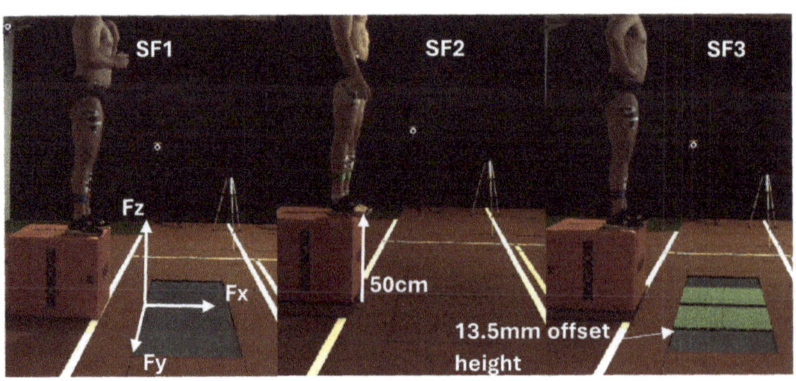

Figure 1. A sagittal view of the participant's starting position before initiating the drop landing, the representation of each flooring surface, and the conventional coordinate system of ground reaction

forces planes (anterior [Fx], lateral [Fy], and vertical [Fz]). SF1 refers to the bare force plate reference condition; SF2 refers to the athletic track condition; and SF3 refers to the interlocked modular flooring tile condition.

2.3. Statistical Analysis

Statistical analyses were performed using jamovi (v2.6.26, jamovi project). Condition-specific individual means for the peak GRFs were calculated using three DL contacts from each condition. Data were screened for normality and sphericity prior to any analysis being conducted. Statistical comparisons were made for surface type (SF1, SF2, and SF3) and limb (Left and Right) using a linear mixed-effects model, with surface type and limbs as fixed effects and participant as a random intercept. Where significant main effects and interactions were present, post-hoc pairwise comparisons were undertaken using a Holm correction. The post-hoc pairwise comparisons were completed between flooring conditions within the same limb (e.g., for vertical GRF; Right limb SF1 vs. Right limb SF2 vs. Right limb SF3) and limb conditions within the same surface (e.g., for vertical GRF; Right limb SF1 vs. Left limb SF1). Effect sizes were reported as partial eta squared (η^2_p) for significant interactions and main effects and Cohen's *dz* for post hoc analyses, which were defined as small (*dz* = 0.20–0.49), medium (*dz* = 0.50–0.79), and large (*dz* ≥ 0.80). All data were reported as means and standard deviations with an alpha level set at 0.05 for all statistical analyses.

3. Results

There were no significant interactions between flooring type and limb for any of the peak GRF variables (Table 1).

Table 1. Peak ground reaction forces across both the left and right lower limb, normalised to body mass, across various flooring types during drop landings from a 50 cm height.

	Bare Force Plate		Athletic Track		Modular Tiles	
Ground Reaction Forces ($N{\cdot}Kg^{-1}$)	**Left**	**Right**	**Left**	**Right**	**Left**	**Right**
[ab] Anterior	6.9 ± 1.1	7.7 ± 1.0	6.7 ± 0.6	7.2 ± 0.6	5.3 ± 0.5	5.7 ± 0.5
Posterior	−2.8 ± 1.1	−3.1 ± 1.0	−2.8 ± 0.9	−3.6 ± 0.7	−2.4 ± 0.6	−2.9 ± 0.9
[b] Lateral	−2.7 ± 0.3	3.5 ± 0.6	−2.8 ± 0.3	4.0 ± 0.5	−2.6 ± 0.4	3.5 ± 0.5
[b] Vertical	24.4 ± 2.9	28.9 ± 3.5	24.4 ± 2.5	30.6 ± 3.1	23.8 ± 2.7	27.4 ± 2.1

'a' represents a significant main effect of flooring type. 'b' represents a significant main effect of limb type.

3.1. Main Effect of Flooring Type

A significant main effect of flooring type was observed in only the peak anterior ($p < 0.001$, $\eta^2_p = 0.221$) GRF. Peak anterior GRF magnitude was smaller in the modular tile condition (mean ± SD; 5.5 ± 0.5 $N{\cdot}Kg^{-1}$) compared with the Mondo (7.0 ± 0.6 $N{\cdot}Kg^{-1}$; $p < 0.001$, $dz = 0.81$) and bare metal force plate conditions (7.3 ± 1.1 $N{\cdot}Kg^{-1}$; $p < 0.001$, $dz = 1.10$). The posterior ($p = 0.313$, $\eta^2_p = 0.024$), lateral ($p = 0.174$, $\eta^2_p = 0.036$), and vertical ($p = 0.588$, $\eta^2_p = 0.011$) GRF directions were not significantly different between flooring types.

3.2. Main Effect of Limb

A significant main effect of limb type was observed in the anterior GRF ($p = 0.048$, $\eta^2_p = 0.040$). There was a smaller peak anterior GRF in the left limb (6.3 ± 0.7 $N{\cdot}Kg^{-1}$; $p = 0.048$, $dz = 0.41$) compared with the right limb across all surfaces (6.9 ± 0.7 $N{\cdot}Kg^{-1}$;

Table 1). The posterior GRF was not significantly different between limbs ($p = 0.066$, $\eta^2_p = 0.035$). There were significant main effects for limb for lateral ($p < 0.001$, $\eta^2_p = 0.280$) and vertical ($p = 0.001$, $\eta^2_p = 0.100$) GRFs. There were smaller peak GRFs in the left limb (Lateral to the left: 2.7 ± 0.3 N·Kg^{-1}, $p < 0.001$, $dz = 1.25$; Vertical: 24.2 ± 2.7 N·Kg^{-1}, $p < 0.001$, $dz = 0.67$) compared with the right limb (Lateral to the right: 3.7 ± 0.5 N·Kg^{-1}; Vertical: 29.0 ± 2.9 N·Kg^{-1}) across all surfaces.

4. Discussion

This study explored the effect of flooring surface on ground reaction forces during bilateral drop landings. Peak anterior GRFs were significantly smaller in the modular tiles compared with the athletic track and bare metal force plate surfaces. Anterior, lateral, and vertical GRFs were all significantly greater in the right limb across all flooring surfaces. The hypothesis was not supported, as peak vertical ground reaction forces were similar across surface types. However, significant differences were observed in the anterior direction, where peak GRFs were smaller in the modular tile condition compared with the athletic sports track and bare metal force plate conditions. Collectively, the results suggest that the flooring surface can affect peak GRFs in the anterior-posterior direction but not in the downwards vertical and lateral directions, with the modular tile surface exhibiting reduced anterior forces during bilateral drop landing tasks.

4.1. Vertical GRF Landing Symmetry

Landing is a standard action performed in sports on various surfaces. Depending on the sport's task requirements, individuals will complete a landing task preceded by a standing or running jump with one to two limbs, commonly reaching heights up to 65 cm in an athletic population [14]. The current study reported that there were no differences between flooring types in vertical GRFs across a drop height of 50 cm; however, significantly greater peak vertical GRFs were observed in the right limb compared with the left limb (Table 1). A similar pattern was also observed for the anterior and lateral GRFs, whereby the right limb was reported to have greater peak GRFs. The observed difference may be in relation to the instructions that the participants received, that being to "step out with your right limb" from the platform.

Task instruction or cuing is an important consideration when performing jumping-landing activities, as previous research has reported significant differences in vertical GRFs when receiving different verbal instructions during drop jumps [15]. Specifically, when participants received the instruction "jump as high as possible" (1A), there were lower vertical GRFs for the first peak and the Relative Strength Index was lower compared with the extended phrase instruction (2B; "jump as high as quickly as possible and during the landing attempt to dampen the impact at ground contact") [15]. Conversely, vertical GRFs for the second peak and flight time were greater in instruction 1A compared with 2B [15]. It is evident that verbal instruction can cause alterations in landing kinetics and performance variables; therefore, it is possible that with the instruction to step with the right leg, participants may have contacted the right limb force plate earlier compared with the left limb, resulting in a reduced amount of time to create a symmetrical landing distribution in the lower extremities. From a biomechanical standpoint, it is possible that the centre of mass was shifted to the right of the midline from the stepping instruction, resulting in a greater proportion to absorb and decelerate the landing by the right limb initially. It is essential to note that the mean differences between flooring conditions that ranged from 0.7 N·Kg^{-1} in the left limb and 3.2 N·Kg^{-1} in the right limb are minor differences that may suggest the flooring surfaces included in this study have no significant effect on drop landing tasks for vertical ground reaction forces; rather, differences may be influenced by the instruction when performing the task.

Other jumping and landing tasks, such as countermovement jumps (CMJs), have demonstrated similar results across different surface types, where peak vertical GRFs were comparable across natural peat soil composition turf, natural loam composition turf, artificial turf, and a bare force plate [5]. On the contrary, during CMJs with varying flooring surface materials (natural grass, indoor rubber, artificial grass, and sand), jump height has been reported to be greater on natural grass and indoor rubber when compared with artificial grass and sand [16]. It might be suggested that different flooring surface materials may have different restitution properties, which could result in altered peak vertical GRFs during jumping-landing tasks. Previous research has shown that the differing densities of playing surfaces can have an altering effect on an individual's muscular force-generating capacity [17]. Although the restitution profile of the flooring surfaces was not measured in the current study, the material and design of the surfaces may be an important consideration for the currently observed results.

4.2. Anterior, Posterior, and Lateral GRFs

Landing with effective movement strategies is crucial for reducing tissue stress and joint loading experience in the lower body. The peak anterior GRF was significantly lower for the modular tile flooring compared with both the bare force plate and athletic sports track in both limbs (Table 1). It is possible that these results are due to the foot landing strategies adopted by the participants. Previous research has demonstrated that a forefoot landing technique during landing tasks results in a significantly greater internal knee adductor moment compared with a rearfoot strategy [18]. The forefoot landing technique has also been shown to decrease flexion at the hip and knee and increase plantarflexion at the ankle compared with a rearfoot technique during drop landing tasks [19]. Alternatively, the landing strategy adopted by the participants may have caused a changed loading rate (force–time curve), resulting in an increased landing duration period and dispersed force. The participants may have performed a rearfoot landing technique during the modular tile flooring condition compared with the bare force plate and athletic track conditions, possibly resulting in a more posterior leaning centre of mass position or increased landing period and therefore reduced anterior GRF during the drop landing task. The foot landing technique was not controlled, nor was the loading rate calculated for this study, and therefore, this cannot be confirmed without further assessment.

The outcomes of the study can be related to the design of the flooring surface, which in turn can affect surface friction, energy storage, and energy loss [5]. As discussed, the coefficient of restitution of the included flooring surfaces was not assessed in this study; however, the surface design of each flooring is different and likely has varied static and dynamic friction coefficients. The friction force, or the sliding force between surfaces, has a strong dependency on the roughness of the surface and the contact pressure with which it is applied [20]. The design of the modular tile is a unique 300 $\times$ 300 mm square that has anchors on each edge for multiple tiles to interlock and a geometric square surface pattern, which in previous research has been shown to change static and dynamic coefficients of friction, suggested through the interlocking joints that disperse the friction forces [21].

In multidirectional sports, common lower-limb injuries include anterior cruciate ligament (ACL) ruptures and lateral ankle sprains [22–24], which may have an increased risk of manifestation if the shoe–surface friction is high [25]. Through the surface design and the possibility of a modified landing strategy, it is possible that the modular tile surface design and the landing strategy collectively contributed to reducing the anterior peak GRF in the modular tile condition. Given that biomechanical responses to landing tasks may vary across sex and age groups [26], it is possible that the peak GRF responses may vary between biological sexes, which is important in the context of foot and ankle injuries, given that female athletes have an increased frequency and severity compared with their male counterparts [27,28].

An important implication of these findings is the potential use of modular flooring tiles as an alternative surface for athletic performance and rehabilitation, particularly following surface friction-related injuries such as ACL ruptures. The observed reduction in anterior GRFs on the modular tile surface might suggest a corresponding decrease in shear joint loading at the knee. This may help mitigate injury risk during landing tasks; however, this cannot be confirmed from the current study. Elevated knee abduction and internal rotation moments, both known to increase ACL strain via the posterior tibial slope, can be moderated through the use of more compliant surfaces [29]. From a practical standpoint, surface selection should be considered an important modifiable factor in both performance and return-to-play environments. Although the majority of GRFs were unaffected by flooring surface in the current study, the reduction in peak anterior GRF observed with modular tiles suggests their potential as an alternative surface during landing tasks where anterior-posterior loading is important.

4.3. Limitations and Future Directions

Although GRFs were the focus of this research, other biomechanical variables, such as joint kinematics and kinetics, were not measured. These parameters could explain the complex coordination that occurs between joint rotations to different sporting surfaces and may help identify compensatory strategies that influence injury risk [30]. The participants were not blinded to the surfaces on which they were landing during data collection. It is possible that the participants adjusted their landing technique across surfaces, attuning to visual stimuli within the surroundings and surfaces of the landing [31].

The protocol involved a controlled drop landing task from a standardised height within a lab environment. While controlled tasks allow for reliable data collection, they do not account for the variability and complexity of sport-specific movements [32,33]. As such, there is merit in exploring future research on surface-specific responses during unanticipated or sport-relevant movement tasks, including cutting, jumping, and pivoting. Lastly, the surfaces utilised in this study were limited to commercially available flooring types with no characterisation beyond nominal labels. Including quantitative assessments of surface mechanical properties, such as energy restitution, coefficient of friction, or force attenuation profiles, would improve external validity and allow for standardised comparisons across future studies [34,35].

5. Conclusions

This study demonstrated that modular flooring tiles consistently resulted in lower anterior GRFs across both limbs at 50 cm heights compared with athletic sports track and bare metal force plate surfaces. These findings may suggest that the design characteristics of modular tiles may reduce anterior shear forces acting on lower body joints, offering potential benefits for injury risk mitigation. Future research should explore how flooring influences joint kinematics and kinetics strategies during more dynamic and sport-specific tasks.

Author Contributions: Conceptualisation, N.A.B., K.J.M., M.D. and A.H.R.; methodology, N.A.B., M.D. and K.J.M.; validation, N.A.B. and K.J.M.; visualisation, N.A.B., K.J.M., M.D., A.J.V. and

A.H.R.; formal analysis, N.A.B.; investigation, N.A.B.; data curation, N.A.B.; writing—original draft preparation, N.A.B.; writing—review and editing, N.A.B., K.J.M., M.D., A.J.V. and A.H.R.; supervision, A.H.R., M.D. and K.J.M.; project administration, N.A.B. All authors have read and agreed to the published version of the manuscript.

Funding: This research was supported by the Australian Government Research Training Program (RTP), administered by La Trobe University.

Institutional Review Board Statement: The study was conducted according to the guidelines of the Declaration of Helsinki and approved by the Institutional Human Ethics Committee of La Trobe University (#HEC21082; date of approval: 13 April 2021).

Informed Consent Statement: Informed consent was obtained from all participants involved in the study.

Data Availability Statement: The original contributions presented in the study are included in the article, and further inquiries can be directed to the corresponding author.

Conflicts of Interest: The authors report no conflicts of interest.

References

1. Stiles, V.; Dixon, S. Sports Surfaces, Biomechanics and Injury. In *The Science and Engineering of Sport Surfaces*; Dixon, S., Fleming, P., James, I., Carre, M., Eds.; Routledge: London, UK, 2015; pp. 70–97.
2. Dragoo, J.L.; Braun, H.J. The effect of playing surface on injury rate: A review of the current literature. *Sports Med.* **2010**, *40*, 981–990. [CrossRef] [PubMed]
3. Williams, S.; Hume, P.A.; Kara, S. A review of football injuries on third and fourth generation artificial turfs compared with natural turf. *Sports Med.* **2011**, *41*, 903–923. [CrossRef]
4. Laver, L.; Kocaoglu, B.; Cole, B.; Arundale, A.J.; Bytomski, J.; Amendola, A. (Eds.) *Basketball Sports Medicine and Science*; Springer Nature: Cham, Switzerland, 2010.
5. Hatfield, D.L.; Murphy, K.M.; Nicoll, J.X.; Sullivan, W.M.; Henderson, J. Effects of different athletic playing surfaces on jump height, force, and power. *J. Strength Cond. Res.* **2019**, *33*, 965–973. [CrossRef]
6. Malisoux, L.; Gette, P.; Urhausen, A.; Bomfim, J.; Theisen, D. Influence of sports flooring and shoes on impact forces and performance during jump tasks. *PLoS ONE* **2017**, *12*, e0186297.
7. Cismaru, I.; Filipașcu, M.; Fotin, A. Wooden flooring—Between present and future. *Pro Ligno* **2015**, *11*, 39–50.
8. Turcas, O.M.; Fotin, A. The evolution and the characteristics of wooden flooring for gym and sport courts. *Pro Ligno* **2017**, *13*, 37–44.
9. Thomson, A.; Whiteley, R.; Bleakley, C. Higher shoe-surface interaction is associated with doubling of lower extremity injury risk in football codes: A systematic review and meta-analysis. *Br. J. Sports Med.* **2015**, *49*, 1245–1252. [CrossRef] [PubMed]
10. Krosshaug, T.; Nakamae, A.; Boden, B.P.; Engebretsen, L.; Smith, G.; Slauterbeck, J.R.; Hewett, T.E.; Bahr, R. Mechanisms of anterior cruciate ligament injury in basketball: Video analysis of 39 cases. *Am. J. Sports Med.* **2007**, *35*, 359–367. [CrossRef]
11. Walker, A. Modelling and analysis of alternative "tile to tile" attachment mechanism designs, for a modular plastic tile sports surface using FEA. *Procedia Eng.* **2016**, *147*, 818–823. [CrossRef]
12. Wang, H.; Zheng, W.; Ma, Y.; Tang, Y. Shock absorption properties of synthetic sports surfaces: A review. *Polym. Adv. Technol.* **2019**, *30*, 2954–2967. [CrossRef]
13. Heebner, N.R.; Rafferty, D.M.; Wohleber, M.F.; Simonson, A.J.; Lovalekar, M.; Reinert, A.; Sell, T.C. Landing kinematics and kinetics at the knee during different landing tasks. *J. Athl. Train.* **2017**, *52*, 1101–1108. [CrossRef] [PubMed]
14. Conceição, F.; Lewis, M.; Lopes, H.; Fonseca, E.M. An evaluation of the accuracy and precision of jump height measurements using different technologies and analytical methods. *Appl. Sci.* **2022**, *12*, 511. [CrossRef]
15. La Greca, S.; Antonacci, G.; Marinelli, S.; Cifelli, P.; Di Giminiani, R. The acute effect of verbal instructions on performance and landing when dropping from different heights: The ground reaction force-time profile of drop vertical jumps in female volleyball athletes. *Front. Sports Act. Living* **2024**, *6*, 1474537. [CrossRef]
16. Arianasab, H.; Mohammadipour, F.; Amiri-Khorasani, M. Comparison of knee joint kinematics during a countermovement jump among different sports surfaces in male soccer players. *Sci. Med. Football* **2017**, *1*, 74–79. [CrossRef]
17. Sassi, A.; Stefanescu, A.; Bosio, A.; Riggio, M.; Rampinini, E. The cost of running on natural grass and artificial turf surfaces. *J. Strength Cond. Res.* **2011**, *25*, 606–611. [CrossRef]

18. Cortes, N.; Morrison, S.; Van Lunen, B.L.; Onate, J.A. Landing technique affects knee loading and position during athletic tasks. *J. Sci. Med. Sport* **2012**, *15*, 175–181. [CrossRef] [PubMed]
19. Cortes, N.; Onate, J.; Abrantes, J.; Gagen, L.; Dowling, E.; Van Lunen, B. Effects of gender and foot-landing techniques on lower extremity kinematics during drop-jump landings. *J. Appl. Biomech.* **2007**, *23*, 289–299. [CrossRef]
20. Voyer, J.; Zehetner, J.; Klien, S.; Ausserer, F.; Velkavrh, I. Production and tribological characterization of tailored laser-induced surface 3D microtextures. *Lubricants* **2019**, *7*, 67. [CrossRef]
21. Berardo, A.; Costagliola, G.; Ghio, S.; Boscardin, M.; Bosia, F.; Pugno, N.M. An experimental-numerical study of the adhesive static and dynamic friction of micro-patterned soft polymer surfaces. *Mater. Des.* **2019**, *181*, 107930. [CrossRef]
22. Dowling, A.V.; Corazza, S.; Chaudhari, A.M.; Andriacchi, T.P. Shoe-surface friction influences movement strategies during a sidestep cutting task: Implications for anterior cruciate ligament injury risk. *Am. J. Sports Med.* **2010**, *38*, 478–485. [CrossRef] [PubMed]
23. Nigg, B.M.; Stefanyshyn, D.J.; Rozitis, A.I.; Mündermann, A. Resultant knee joint moments for lateral movement tasks on sliding and non-sliding sport surfaces. *J. Sports Sci.* **2009**, *27*, 427–435. [CrossRef] [PubMed]
24. Frias Bocanegra, J.M.; Fong, D.T. Playing surface traction influences movement strategies during a sidestep cutting task in futsal: Implications for ankle performance and sprain injury risk. *Sports Biomech.* **2022**, *21*, 380–390. [CrossRef]
25. Dowling, A.V.; Andriacchi, T.P. The Role of Shoe-Surface Interaction and Noncontact ACL Injuries. In *ACL Injuries in the Female Athlete: Causes, Impacts, and Conditioning Programs*; Springer: Berlin/Heidelberg, Germany, 2018; pp. 97–118.
26. Norcross, M.F.; Johnson, S.T.; Bovbjerg, V.E.; Koester, M.C.; Hoffman, M.A. Factors influencing high school coaches' adoption of injury prevention programs. *J. Sci. Med. Sport* **2016**, *16*, 426–430. [CrossRef]
27. Talia, A.J.; Busuttil, N.A.; Kendal, A.R.; Brown, R. Gender differences in foot and ankle sporting injuries: A systematic literature review. *Foot* **2024**, *60*, 102122. [CrossRef]
28. Talia, A.J.; Busuttil, N.A.; Hotchen, A.; Kendal, A.R.; Brown, R. Sex Differences in Foot and Ankle Sports Injury Rates in Elite Athletes: A Systematic Review and Meta-analysis of 25,687,866 Athlete Exposures. *Orthop. J. Sports Med.* **2025**, *13*, 23259671251364261. [CrossRef]
29. Navacchia, A.; Bates, N.A.; Schilaty, N.D.; Krych, A.J.; Hewett, T.E. Knee abduction and internal rotation moments increase ACL force during landing through the posterior slope of the tibia. *J. Orthop. Res.* **2019**, *37*, 1730–1742. [CrossRef]
30. Wright, I.C.; Neptune, R.R.; van den Bogert, A.J.; Nigg, B.M. The influence of foot positioning on ankle sprains. *J. Biomech.* **2000**, *33*, 513–519. [CrossRef]
31. Liebermann, D.G.; Goodman, D. Effects of visual guidance on the reduction of impacts during landings. *Ergonomics* **1991**, *34*, 1399–1406. [CrossRef]
32. McNitt-Gray, J.L. Kinetics of the lower extremities during drop landings from three heights. *J. Biomech.* **1993**, *26*, 1037–1046. [CrossRef] [PubMed]
33. Donelon, T.A.; Edwards, J.; Brown, M.; Jones, P.A.; O'Driscoll, J.; Dos' Santos, T. Differences in biomechanical determinants of ACL injury risk in change of direction tasks between males and females: A systematic review and meta-analysis. *Sports Med.-Open* **2024**, *10*, 29. [CrossRef] [PubMed]
34. Nigg, B.M.; Liu, W. The effect of muscle stiffness and damping on simulated impact force peaks during running. *J. Biomech.* **1999**, *32*, 849–856. [CrossRef] [PubMed]
35. Dixon, S.J.; Collop, A.C.; Batt, M.E. Surface effects on ground reaction forces and lower extremity kinematics in running. *Med. Sci. Sports Exerc.* **2000**, *32*, 1919–1926. [CrossRef] [PubMed]

symmetry

MDPI

Article

Laterality, Shot Direction and Spatial Asymmetry in Decisive Penalty Kicks: Evidence from Elite Men's Football

Pablo Cidre-Fuentes, Manuel Alberto González-Harcevnicow and Iván Prieto-Lage *

Observational Research Group, Faculty of Education and Sport, University of Vigo, 36005 Pontevedra, Spain; pablojose.cidre@uvigo.gal (P.C.-F.); manuel.gonzalez.harcevnicow@alumnado.uvigo.gal (M.A.G.-H.)
* Correspondence: ivanprieto@uvigo.es

Abstract: Penalty shootouts often decide major football tournaments, making the analysis of spatial symmetry and shot patterns crucial for performance optimization. This study analyzed 212 decisive penalty kicks in elite men's football to explore spatial patterns and asymmetries in execution, as well as their relationship with performance effectiveness. An observational methodology was used, combining temporal pattern detection (T-patterns) and chi-square tests to examine associations between contextual, spatial, and outcome-related variables. Results showed that the most frequently targeted area was left-down (28.3%), with a success rate of 71.7%. Additionally, central zones exhibited particularly high accuracy (ranging from 88.9% to 100%) despite their low usage frequency. Differences were also observed in the distribution of shots between left- and right-footed players, both in frequency and effectiveness, although these were not significant. The findings suggest the presence of strategic tendencies and functional spatial asymmetries, which may have implications for specialized training in high-pressure scenarios. These insights can guide targeted training strategies for both kickers and goalkeepers and encourage further research on decision-making and spatial behavior under extreme pressure.

Keywords: spatial symmetry; systematic observation; performance analysis; T-patterns; decision-making under pressure; key performance indicators

1. Introduction

The execution of penalty kicks in elite football represents a critical event, often determining the outcome of high-stakes matches or entire competitions [1]. It is a specific action in the game in which only two players are directly involved: the kicker and the goalkeeper [2]. This binary interaction, framed within a spatially symmetric environment (the goal structure), invites detailed analysis of how players exploit or break that symmetry to gain an advantage [3]. In this study, the term decisive penalty refers exclusively to a penalty kick taken during a shootout whose outcome directly determines the result of the match—either by securing immediate victory for the kicker's team or by causing their elimination if missed. Penalties taken earlier in the shootout, which do not yet mathematically decide the outcome, are not included within this definition.

The study of penalties in football has become an important part of performance prediction and behavioral analysis in sports [1,4,5]. These situations are highly advantageous for the kicker, who faces no opposition and therefore typically achieve success rates of 69–85% [2,6–10]. Performance does not usually differ between home and away contexts [10–12].

Penalty kicks follow a series of trends that have been identified in various studies [2,8–14]. These trends are influenced by subtle spatial decisions, lateral preferences, and psychological pressure, particularly factors such as the high responsibility of decisive kicks, the potential consequences of elimination, or the increased anxiety associated with taking the final shots in a shootout [15,16]. Generally, penalties tend to be taken to the sides of the goal, more specifically, to the right of the goalkeeper [2,8,14], accounting for 34.7–39.7% of the total [8,13,14]. This tendency is more pronounced among right-footed players, who tend to shoot to the right 59–62% of the time—a direction identified by several studies as their 'natural' side [2,13,17]. In contrast, left-footed players display greater variability, with right-sided shooting occurring in 32–59% of cases, depending on the category analyzed [2].

The interaction between the goalkeeper and the penalty taker has been examined in various studies, with particular attention to the goalkeeper's movements prior to the execution of the penalty kick [5,7,8,18]. Some authors suggest that such anticipatory movements may influence the decision-making process of the penalty taker [5], whereas another study indicates that this strategy might not be effective at the elite level of football [19]. Furthermore, in attempts to predict the direction of the shot, certain biomechanical indicators—such as the orientation of the supporting foot and the position of the arm contralateral to the kicking leg—have been proposed as reliable cues in some investigations [5].

In elite international tournaments—both club-level and national—the role of penalty shootouts becomes particularly critical in knockout stages. After extra time, if the score remains tied, the match is resolved by a series of kicks that not only test physical execution, but also decision-making under intense psychological stress [20]. Finalists in the World Cup have a 56.5% chance of being involved in at least one penalty shootout; this percentage is slightly lower for European teams, at 46.5% [21].

The order in which penalty shots are taken also seems to influence the outcome of a shootout. Around 59.2–60.5% of victories go to the team that starts by taking the first shot [22].

There are different theories about the order of the kickers in these shootouts. Some indicate that players should take penalties in reverse order of ability, meaning the best shooter should take the fifth kick. However, probability theory suggests that the last penalties in a shootout are the least likely to succeed [16]. If the score remains tied after the first five throws, the success rate drops to 64.3% [16]. This may be because the importance of these kicks negatively affects the players' performance, due to the stress of taking these penalties. Psychological components constitute a significant determinant in the success or failure of penalty kicks [15].

These moments of pressure can disrupt regular behavioral patterns [23], producing identifiable symmetry-breaking tendencies—whether in shot placement, body orientation, or outcome probabilities. Understanding these patterns provides valuable insight into motor control, anticipation, and decision-making under constraints, all of which are key components in the study of symmetry in sports contexts.

Identifying these patterns is crucial for informing tactical preparation and optimizing performance under high-pressure conditions. Given the importance of these situations in determining tournament outcomes, this study aims to analyze and detect spatial and sequential patterns in decisive penalty kicks (as defined above) from shootouts in elite international football competitions between 2010 and 2023. Through observational methodology, statistical analysis, and T-pattern detection, this research explores how symmetry—both spatial and behavioral—is maintained or broken in these high-pressure scenarios, offering practical applications for performance optimization in elite sport. In line with these aims, we hypothesized that decisive penalty kicks would show structured asymmetries in both spatial distribution and sequential patterns, reflecting adaptive responses to psychological pressure. Unlike general penalty situations, decisive shootout kicks are expected to amplify motor and tactical regularities, making them particularly suitable for systematic analysis. Previous research has mostly examined penalties in broader contexts (e.g., regular match penalties or aggregated shootout data), leaving a gap in the understanding of the specific dynamics of decisive penalties. By combining observational methodology with T-pattern detection, this study addresses this gap and contributes to the literature by uncovering how symmetry is strategically maintained or broken under the highest levels of competitive stress. Addressing this gap is important not only for advancing theoretical knowledge on motor behavior under pressure, but also for providing applied insights to optimize performance preparation in elite football.

2. Materials and Methods

2.1. Design

An observational methodology [24] was employed to analyze the execution of decisive penalties in penalty shootouts that occurred between 2010 and 2023. This methodological approach involves the systematic, non-participant observation of naturally occurring behaviors in their real competitive context, recorded and coded according to a predefined category system. The observational design [24] was nomothetic (each penalty was analyzed independently), follow-up (spanning the period from 2010 to 2023), and unidimensional (one level of response).

2.2. Sample

The sample consisted of all decisive penalties from penalty shootouts in both club (UEFA Champions League, UEFA Europa League and UEFA Conference League) and national team matches (FIFA World Cup, UEFA Euro, UEFA Nations League and CONMEBOL Copa América), which took place in continental and world competitions between 2010 and 2023 (66 penalty shootouts). A total of 212 decisive penalties were analyzed. Inclusion criteria were: (a) penalty kicks taken in shootouts of elite-level continental or world competitions, involving either national teams or European club teams; (b) penalties meeting our definition of "decisive" as defined in the Introduction. Exclusion criteria were: (a) penalties from shootouts not meeting the "decisive" definition; (b) penalties from competitions outside the specified tournaments. All available decisive penalties meeting this definition were included, with no cases excluded due to video quality, ambiguous outcomes, duplicate footage, or annulled decisions.

The study was approved by the ethics committee of the Faculty of Education and Sport Science (University of Vigo, application 03-090425).

2.3. Instruments

The instrument described in Table 1 was developed ad hoc for this study and is a category system in which the criteria categories are mutually exclusive. This instrument is composed of nine criteria, which are made up of a total of thirty-two categories. The criteria and categories were established based on previous research with similar objectives [10,11] and were subsequently refined through consultation with two experts in football and observational methodology. The construct validity of the instrument was assessed by verifying its consistency with the theoretical framework [24] and by obtaining expert agreement on the adequacy of the categories and definitions, achieving a satisfaction rate of 95%.

Table 1. Observational instrument, descriptive statistics, and results of chi-square goodness-of-fit and independence tests for each criterion.

Criteria	Description	*n*	%	Ef.	χ^2 G-O-F	χ^2 Indep.
Type of match	Club match	97	45.8	73.2	$\chi^2 = 1.528$ $p = 0.216$	$\chi^2 = 2.686$ $p = 0.101$
	National team match	115	54.2	62.6		
Stadium	The team that kicks plays as the home team	35	16.5	77.1	$\chi^2 = 105.009$ $p < 0.001$	$\chi^2 = 2.196$ $p = 0.334$
	The team that kicks plays as the away team	36	17.0	61.1		
	Not-applicable. Neutral venue	141	66.5	66.7		
Round	Round of 32 match	30	14.2	63.3	$\chi^2 = 31.066$ $p < 0.001$	$\chi^2 = 1.470$ $p = 0.832$
	Round of 16 match	42	19.8	71.4		
	Quarterfinal match	71	33.5	63.4		
	Semifinal match	24	11.3	70.8		
	Final match	45	21.2	71.1		
Type of Penalty	If the kicker misses, their team lose the shootout	57	26.9	68.4	$\chi^2 = 12.632$ $p = 0.002$	$\chi^2 = 9.499$ $p = 0.009$
	If the kicker scores, their team wins the shootout	60	28.3	81.7		
	If the kicker misses, the next team can win if they scores the penalty kick	95	44.8	57.9		
Order	The team takes the first penalty in the shootout	100	47.2	71.0	$\chi^2 = 0.679$ $p = 0.410$	$\chi^2 = 1.085$ $p = 0.298$
	The team does not take the first penalty in the shootout	112	52.8	64.3		
Laterality	The kicking player is right-footed	163	76.9	65.6	$\chi^2 = 61.302$ $p < 0.001$	$\chi^2 = 1.051$ $p = 0.305$
	The kicking player is left-footed	49	23.1	73.5		
Run-up to the Kick	The kicker takes fewer than three steps before striking the ball	206	97.2	67.0	$\chi^2 = 388.877$ $p < 0.001$	$\chi^2 = 4.505$ $p = 0.105$
	The kicker takes three or more steps before striking the ball	5	2.4	100		
	The kicker clearly pauses during the run-up	1	0.5	0		
Direction-Goal (depending on the view of the kicker)	Left-Top	28	13.2	60.7	$\chi^2 = 105.377$ $p < 0.001$	$\chi^2 = 17.128$ $p = 0.029$
	Left-Medium height	21	9.9	47.6		
	Left-Down	60	28.3	71.7		
	Centre-Top	6	2.8	100		
	Centre-Medium height	9	4.2	88.9		
	Centre-Down	9	4.2	88.9		
	Right-Top	17	8.0	52.9		
	Right-Medium height	20	9.4	85.0		
	Right-Down	42	19.8	59.5		
Direction-Laterality (based on near or far post with respect to foot dominance)	Kicker-top	31	14.6	58.1	$\chi^2 = 116.670$ $p < 0.001$	$\chi^2 = 11.372$ $p = 0.181$
	Kicker-medium	29	13.7	62.1		
	Kicker-down	61	28.8	72.1		
	Middle-top	6	2.8	100.0		
	Middle-medium	9	4.2	88.9		
	Middle-down	9	4.2	88.9		
	Far-top	14	6.6	57.1		
	Far-medium	12	5.7	75.0		
	Far-down	41	19.3	58.5		
Effectiveness	The penalty ends in goal	143	67.5	-	$\chi^2 = 112.255$ $p < 0.001$	-
	The penalty is saved by the goalkeeper	41	19.3			
	The penalty is missed (off target)	28	13.2			

Note: Ef.: Effectiveness; G-O-F: Goodness-of-fit; Indep.: Independence.

Data recording was carried out using LINCE PLUS software version 2.1 (Alberto Soto Fernández, INEFC, Lleida, Spain) [25]. This interactive and highly flexible multimedia tool is specifically designed to facilitate systematic observation by enabling the analysis and recording of sports actions. The software allows for the configuration of buttons on the computer interface corresponding to the codes previously assigned to each category within the observation instrument. By generating an observational record of specific actions during sporting events, it contributes to a deeper understanding of behavioral patterns in sports performance. LINCE PLUS has been widely used in numerous studies employing systematic observational methodology in sports settings (e.g., [4,26–28]), which supports its suitability for the present research.

The full set of criteria and operational definitions used in the observational tool will be detailed in the Table 1 and Figure 1.

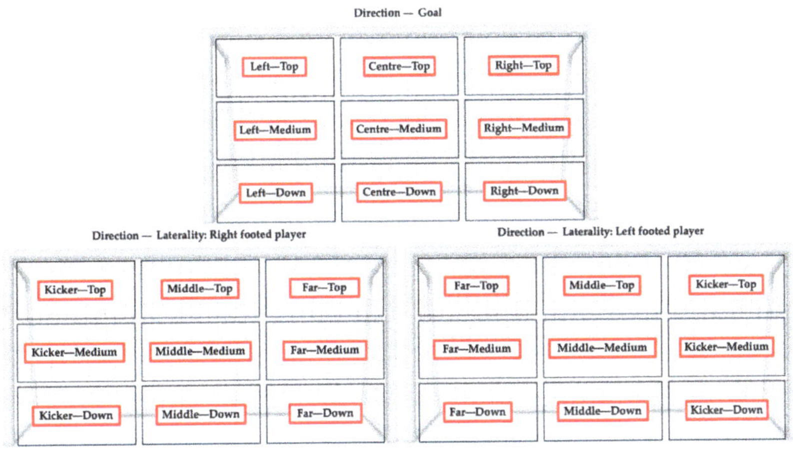

Figure 1. Goal areas according to direction-goal and direction-laterality.

2.4. Procedure

The penalty kicks analyzed in this study were obtained from publicly available sources, including YouTube and the official platforms of UEFA (https://www.youtube.com/@UEFA, accessed on 15 September 2024) and FIFA (https://www.youtube.com/@fifa, accessed on 15 September 2024). Subsequently, the video footage was edited to produce two separate files—one for club teams and another for national teams—in which the penalties were organized chronologically by match day. The editing process was carried out using iMovie software (version 10.4.3; Apple Inc., Cupertino, CA, USA). This approach facilitated the creation of user-friendly video files optimized for use with the observational recording tool.

After conducting training in the use of the instruments, and in order to ensure methodological rigor in the coding process [24], a quality control procedure was implemented to verify the reliability of the recorded data. This was carried out through the calculation of both intra- and inter-observer agreement using Cohen's Kappa coefficient (COH), performed using LINCE Plus. To assess intra-observer reliability, a concordance test was conducted on 25% of the total sample of penalty kicks. The analysis yielded a Kappa coefficient of 0.99 for both the first and second observations, indicating an almost perfect level of agreement. Inter-observer reliability was then evaluated using the same subset of data, also resulting in a Kappa coefficient of 0.99. These results confirm the robustness and consistency of the observational process. Once the quality of observation was verified, the full dataset was systematically coded by a single trained observer.

Following the completion of the penalty recordings, the data were exported to an Excel file. This file provided a detailed account of the sequential order of the recorded behavior codes, along with their temporal occurrence and corresponding duration, expressed in frames. The versatility of the exported dataset allowed for subsequent transformations, enabling its adaptation for various types of analytical procedures [29].

2.5. Data Analysis

Descriptive and inferential statistical analyses were conducted using IBM SPSS Statistics, version 25.0 (IBM-SPSS Inc., Chicago, IL, USA). A chi-square goodness-of-fit test was applied to determine whether the distribution of observations within each category differed significantly from a uniform distribution, thereby highlighting patterns of preference or dominance in each variable. Furthermore, chi-square tests of independence were carried out to assess whether the effectiveness of the penalty kick (i.e., goal, save, or miss) was significantly associated with the categories of the other criteria. These statistical procedures provide a comprehensive overview of both intra- and inter-criterion relationships, offering an initial understanding of the structural dynamics underlying decisive penalty kicks. Statistical significance was assumed for $p < 0.05$.

To examine penalty patterns, temporal pattern detection was conducted using THEME v.6.0 (PatternVision Ltd., Reykjavík, Iceland) [30]. THEME has been extensively applied and validated in systematic observational research, particularly in sports performance contexts (e.g., [5,31]). This software is designed to detect hidden temporal structures (T-patterns) within behavioral data, allowing researchers to identify recurrent sequential patterns that are not accessible through conventional statistical analyses. Its application in the present study makes it possible to reveal underlying spatiotemporal strategies in decisive penalty kicks, thereby contributing novel insights into performance under high-pressure conditions.

3. Results

3.1. Descriptive and Inferential Statistical Outcomes

To begin the presentation of results, Table 1 shows the observational instrument used in this study, along with the descriptive statistics for each criterion.

The analysis revealed significant differences in the distribution of frequencies across most observational criteria. Only "type of match" ($p = 0.216$) and "order of the kick" ($p = 0.410$) showed no significant deviations from a uniform distribution, suggesting that choices in these two cases were more evenly balanced.

In contrast, "penalty effectiveness" was significantly associated with both "type of penalty kick" ($p = 0.009$) and "goal direction" ($p = 0.029$), indicating that the situational context of the kick (e.g., immediate victory or elimination) and the target area selected by the kicker have a decisive influence on the outcome.

Regarding success rates, penalties taken while playing at home showed the highest conversion rate (77.1%), followed by those taken on neutral grounds (66.7%), while penalties executed as the away team had the lowest success rate (61.1%). Kicks that provided the opportunity to win the match if scored achieved the highest effectiveness (81.7%), whereas those where a miss would result in immediate elimination recorded the lowest conversion rate (57.9%).

From a spatial perspective, over half of the penalty kicks (52.3%) were aimed at the lower zones of the goal, both central and lateral, aligning with the general tendency to target low areas to reduce the risk of missing. Overall, 67.5% of the penalties were successfully converted, 19.3% were saved by the goalkeeper, and 13.2% were missed entirely (off-target or hitting the post).

Concerning the relationship between shot direction and player laterality, the data showed a higher frequency of kicks directed toward the dominant-foot side of the player, known as the near post. The zones classified as "kicker side" accounted for 14.6%,

13.7%, and 28.8% of the shots, whereas the "far post" zones registered lower frequencies (6.6%, 5.7%, and 19.3%). Additionally, left-footed players achieved a success rate of 73.5%, compared to 65.6% for right-footed players. However, this difference was not significant ($p = 0.584$).

Figure 2 displays the effectiveness of penalty kicks based on the area of the goal targeted, without distinguishing between left- or right-footed players. The left panel includes all penalty kicks taken ($n = 212$), while the right panel focuses only on those directed between the goalposts ($n = 184$), meaning shots that were either saved or scored.

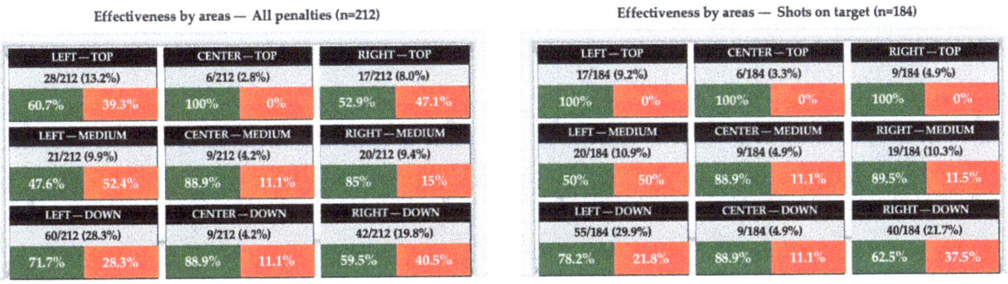

Figure 2. Effectiveness by areas direction-goal (all penalties) and effectiveness by areas direction-goal (shots on target).

In terms of usage frequency ($n = 212$), the most commonly targeted areas were left-down (28.3%), right-down (19.8%), and left-top (13.2%), which together accounted for over 60% of all penalty kicks. However, when analyzing only shots directed on target ($n = 184$), their effectiveness varied: left-down achieved a success rate of 78.2%, right-down 62.5%, and left-top 100%. In contrast, other frequently used zones such as left-medium (9.9% of total penalties) recorded a lower success rate, with only 50% of on-target attempts resulting in goals.

Other less frequently used zones also exhibited high performance. Right-top, for instance, accounted for just 8.0% of total kicks but achieved 100% success on shots on target ($n = 9$). The same was true for the center-top area (2.8% of total kicks), which was the only zone to reach perfect effectiveness both in overall attempts and in shots directed on target. The center-middle and center-down areas also showed high success rates, each converting 88.9% of on-target shots.

Overall, the central areas of the goal, especially center-top, demonstrated the highest levels of accuracy. Despite their low usage, their performance indicates that they may be technically effective target options in decisive penalty scenarios.

Figure 3 shows the effectiveness of penalty kicks based on the targeted area of the goal, categorized according to the direction of the shot relative to the kicker's dominant foot: kicker side (short post), middle, and far side (long post). The left panel includes all penalties taken ($n = 212$), while the right panel considers only those directed on target ($n = 184$), meaning shots that resulted in a goal or were saved by the goalkeeper.

This reclassification by laterality allows for a more precise analysis of tendencies between short and long post targeting. The most frequently used zone was kicker-down (28.8%), followed by far-down (19.3%) and kicker-top (14.6%). Kicker-down achieved a success rate of 72.1% for all kicks and 78.6% for on-target attempts. In contrast, far-down showed lower success: 58.5% overall and 61.5% on target.

In the upper zones, kicker-top recorded an overall success rate of 58.1%, while far-top showed 57.1%, with both areas presenting a relatively high proportion of missed or saved attempts. Regarding the middle-height areas, kicker-medium achieved 62.1% effectiveness on target shots, compared to 75.0% for far-medium, although the latter was used less frequently.

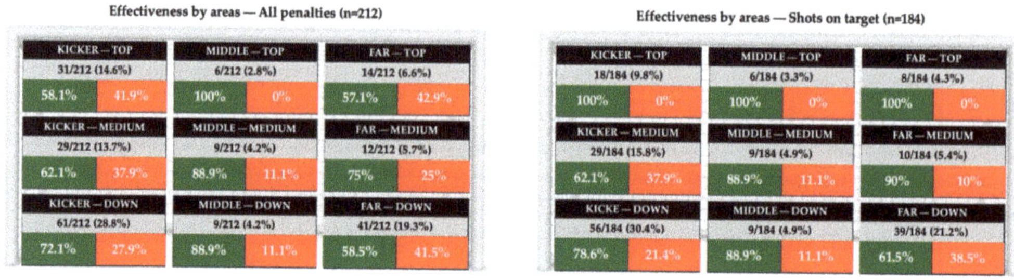

Figure 3. Effectiveness by areas direction-laterality (all penalties) and effectiveness by areas direction-laterality (shots on target). Note: "Kicker side" refers to the near post relative to the kicker's dominant foot, and "far side" refers to the opposite post. These spatial references apply equally to both right-footed and left-footed players.

Overall, kicker-side zones were more frequently selected and generally yielded higher success rates compared to far-side zones, which may reflect a biomechanical or perceptual preference under high-pressure conditions.

Figure 4 shows the effectiveness of penalty kicks based on shot direction and the kicker's laterality, distinguishing between right-footed and left-footed players. As in Figure 3, the left panels include all penalties (n = 212), while the right panels consider only on-target shots (n = 184).

Figure 4. Effectiveness by areas direction-laterality (**right** and **left**) and effectiveness by areas direction-laterality (**right** and **left**). Note: The top row corresponds to right-footed kickers, while the bottom row shows the performance of left-footed kickers.

Among right-footed players, the most frequently used zones were kicker-down (30.1%), far-down (18.4%), and kicker-top (16%). When considering only shots on target, success rates increased: kicker-down went from 69.4% to 77.3%, far-down from 50.0% to 53.6%, and kicker-top from 61.5% to 100%. The middle-top and middle-medium zones reached 100% accuracy in both conditions, while far-medium rose from 90.0% to 100%, and middle-down remained stable at 83.3%.

For left-footed players, the most commonly used zones were kicker-down (24.5%), far-down (22.4%), and kicker-medium (20.4%). The most effective zone was middle-down, with 100% accuracy in both total attempts and on-target shots. Far-top and kicker-top also achieved 100% success on shots on target, although their overall success rates were lower—50% and 40.0%, respectively—due to off-target attempts. Other effective areas included kicker-down (83.3%), far-down (81.8%), and middle-medium (75.0%). Far-medium did not record any successful penalties (0.0%).

When considering only shots on target, right-footed players stood out in middle-top, middle-medium, far-top, far-medium, and kicker-top, all with 100% success. They also achieved good results in middle-down (83.3%) and kicker-down (77.3%). Lower success rates were observed in kicker-medium (52.6%) and far-down (53.6%).

Left-footed players, in turn, achieved 100% success in middle-down, far-top, and kicker-top, and high percentages in kicker-down (83.3%), far-down (81.8%), and middle-medium (75.0%). Far-medium was the only zone without any goals on target.

In terms of spatial distribution, right-footed players directed 57.8% of their penalties toward their dominant side (kicker side), while left-footed players did so in 55.1% of the cases.

3.2. Analysis of Decisive Penalty Kick T-Patterns

To complement the descriptive and inferential analyses, a T-pattern detection procedure was conducted in order to explore the temporal structure and contextual regularities underlying decisive penalty kicks. This analysis was performed using the THEME software and focused exclusively on patterns with a minimum of three occurrences.

A total of 2435 unique T-patterns were identified and classified based on the type of decisive penalty kick:

- Miss—Lose (penalties missed that directly resulted in a loss).
- Score—Win (successful penalties that secured victory).
- Miss—Continue (penalties missed that allowed the opponent to continue and potentially win).

Each category was then subdivided according to the kicker's laterality (left-footed or right-footed) and the outcome of the kick (goal, save, or off-target).

Due to the volume and complexity of the data, the full results are provided as Supplementary Material, including three detailed tables:

- Table S1 presents the most frequent T-patterns in Miss-Lose situations, revealing key factors such as run-up characteristics, shot direction, and contextual variables (e.g., venue, round).
- Table S2 includes the T-patterns associated with Score-Win penalties, highlighting conditions that consistently precede successful outcomes.
- Table S3 covers the patterns observed in Miss-Continue situations, offering insight into behaviors that may increase risk or reduce effectiveness.

This structured temporal analysis enriches the understanding of behavioral strategies adopted by players under high-pressure conditions and complements the static performance indicators presented earlier in the article.

4. Discussion

4.1. General Overview and Key Findings Under Pressure

Decisive penalty kicks in elite football represent unique high-pressure scenarios where psychological, tactical, and motor control factors interact to determine success or failure. Our findings reinforce previous research suggesting that performance under these conditions is shaped by both the situational context and the cumulative effects of stress during the shootout [6,10,12]. The advantage observed for teams taking the first kick aligns with earlier studies reporting similar trends [22], supporting the view that initiating the sequence can confer a psychological edge by placing immediate pressure on the opponent.

Reduced effectiveness in the final kicks of the sequence is consistent with the idea that rising pressure impairs technical execution [16,22]. This phenomenon may relate to the "choking under pressure" effect described in sports psychology literature [32]. whereby heightened anxiety and perceived responsibility disrupt motor performance. Moreover, the decisive nature of these kicks likely increases cognitive load, affecting attentional focus and decision-making in ways not observed in non-decisive penalties [8,9,31]. Together, these findings highlight the need to examine both spatial and temporal patterns to understand adaptive responses under pressure and to develop targeted preparation strategies.

4.2. Spatial Direction and Zone Preference

The tendency to target lateral areas, particularly the goalkeeper's right, is consistent with previous evidence on laterality effects in penalty taking [2,8,10,11,14]. This bias may stem from biomechanical preferences—especially among right-footed players—linked to natural kicking mechanics and the perception–action coupling described in motor control theories [2,8,10]. However, our findings suggest that high-pressure contexts can disrupt these regularities, producing symmetry-breaking behaviors such as targeting the central vertical corridor. Although less frequent, central shots—particularly to the middle-top and middle-medium zones—showed notably high success rates, which may reflect opportunistic exploitation of goalkeepers' anticipatory lateral movements [11,14].

From a theoretical perspective, the intentional breaking of spatial symmetry can be interpreted through established frameworks in motor control, decision-making under stress, and ecological dynamics. Under pressure, athletes often revert to well-learned motor patterns that offer proprioceptive control and minimize execution variability, consistent with Schmidt's Schema Theory [33] and optimal control models [34]. Ecological dynamics emphasizes that such behaviors emerge from the interaction between task, environmental, and individual constraints, with performers exploiting affordances that maximize perceived control [35]. Attentional Control Theory further posits that anxiety shifts focus toward goal-relevant cues, leading athletes to select motor solutions that feel most reliable [32]. Thus, symmetry-breaking in penalty kicks appears to be a functional adaptation to the demands of the situation rather than random variation.

4.3. Laterality and Footedness

Differences between right- and left-footed players in shot distribution and accuracy, although not statistically significant, follow patterns reported in earlier studies [2,8,10]. Right-footed players showed a stronger bias toward their dominant side, while left-footed players displayed greater variability, supporting the view that minority laterality profiles introduce unpredictability and challenge goalkeeper anticipation [2,8,10]. From a performance standpoint, these findings underline the value of individualized penalty preparation that accounts for the kicker's laterality and the goalkeeper's tendencies. Coaches could leverage the variability of left-footed players in decisive contexts, as their unpredictability may confer a tactical advantage.

4.4. Temporal Pattern Analysis (T-Patterns)

The use of T-pattern detection revealed sequential structures in penalty execution that are not apparent in conventional statistics. These patterns, consistent with prior applications of THEME in sports performance analysis [4,31], indicate that certain spatial choices and outcomes cluster under specific contextual conditions. Such structures align with the concept of emergent coordination patterns in ecological dynamics, where athletes adapt based on environmental cues and the evolving sequence of play.

By uncovering these hidden temporal regularities, our study adds to the evidence supporting T-pattern analysis as a tool for revealing the underlying logic of performance behaviors. This approach offers practical value by enabling coaches to anticipate and train for recurrent behavioral sequences in high-pressure scenarios.

4.5. Contributions and Applied Implications

This research provides new evidence on how spatial symmetry is maintained or broken during decisive penalty kicks, integrating both spatial and temporal perspectives. By highlighting the interplay between biomechanical preferences, psychological pressure, and sequential dependencies, the findings can inform targeted training interventions for both kickers and goalkeepers. For example, goalkeepers may benefit from training to resist anticipatory movements that open central shooting lanes, while kickers can be trained to exploit these openings effectively.

Moreover, the methodological integration of systematic observation with T-pattern detection offers a replicable framework for analyzing performance behaviors in other sports contexts. The dual focus on spatial and temporal patterns enriches the tactical understanding of penalty shootouts, providing actionable insights for elite-level preparation.

4.6. Limitations and Future Research Directions

This study focused exclusively on decisive penalty kicks in elite international competitions, which may limit its applicability to league matches or non-decisive penalties. The relatively small number of left-footed kickers restricted the robustness of laterality-based comparisons, and goalkeeper anticipation or movement patterns—despite their potential influence—were not examined. The T-pattern methodology is sensitive to coding quality and temporal resolution; integrating multimodal data (e.g., biomechanics, gaze tracking, physiological indicators) could strengthen analyses. The predefined goal-zone grid could also be replaced or complemented by dynamic spatial models, such as ball-trajectory heatmaps or AI-based goalkeeper tracking.

Future studies should test training interventions to manage spatial–temporal asymmetries, analyze decision-making under fatigue or emotional stress, and explore kicker–goalkeeper interaction dynamics to enhance performance in high-pressure contexts.

5. Conclusions

This study offers new insights into how elite football players respond to high-pressure situations by selectively breaking spatial and behavioral symmetry during decisive penalty kicks. The results confirm that penalty shootouts are not purely stochastic events but are governed by strategic tendencies shaped by biomechanical familiarity, perceptual comfort, and psychological stress.

Despite the theoretically symmetrical nature of the goal structure, kickers consistently favored the lower zones—particularly the short-post areas on their dominant side—suggesting a functional asymmetry driven by motor control strategies. Interestingly, central areas, although infrequently targeted, demonstrated the highest success rates, pointing to untapped opportunities that remain underutilized in practice.

Contextual factors such as venue, order of execution, and type of penalty (i.e., win/loss consequence) also influenced outcomes, highlighting the role of situational asymmetries beyond spatial configuration. These findings suggest that both environmental and internal (motor/perceptual) asymmetries must be considered when analyzing performance under pressure.

Moreover, the T-pattern analysis revealed structured behavioral sequences that distinguish successful from unsuccessful kicks, underscoring the importance of temporal dynamics and routine consistency in high-stakes contexts.

From an applied perspective, these results support tailored training approaches that incorporate not only spatial variability in shot placement but also psychological resilience and decision-making under stress. For coaches and analysts, understanding the nuanced interplay between symmetry, laterality, and outcome can inform more effective strategies in preparation for penalty shootouts.

Supplementary Materials: The following supporting information can be downloaded at: https://www.mdpi.com/article/10.3390/sym17091570/s1, Table S1: T-Patterns in Miss-Lose Penalties; Table S2: T-Patterns in Score-Win Penalties; Table S3: T-Patterns in Miss-Continue Penalties.

Author Contributions: Conceptualization, P.C.-F., M.A.G.-H. and I.P.-L.; methodology, P.C.-F., M.A.G.-H. and I.P.-L.; software, P.C.-F., M.A.G.-H. and I.P.-L.; validation, P.C.-F., M.A.G.-H. and I.P.-L.; formal analysis, P.C.-F., M.A.G.-H. and I.P.-L.; investigation, P.C.-F., M.A.G.-H. and I.P.-L.; resources, P.C.-F., M.A.G.-H. and I.P.-L.; data curation, P.C.-F., M.A.G.-H. and I.P.-L.; writing—original draft preparation, P.C.-F., M.A.G.-H. and I.P.-L.; writing—review and editing, P.C.-F., M.A.G.-H. and I.P.-L.; visualization, P.C.-F., M.A.G.-H. and I.P.-L.; supervision, P.C.-F., M.A.G.-H. and I.P.-L.; project administration, P.C.-F., M.A.G.-H. and I.P.-L.; funding acquisition, I.P.-L. All authors have read and agreed to the published version of the manuscript.

Funding: This study was funded by the Ministerio de Ciencia, Innovación y Universidades (MCIU), the Agencia Estatal de Investigación (AEI), and the European Union (EU) under the Project LINCE PLUS: Multimodal platform for data integration, synchronization and analysis in physical activity and sport [PID2024-156051NB-I00] (2025–2027), awarded to Iván Prieto-Lage.

Institutional Review Board Statement: The study was approved by the ethics committee of the Faculty of Education and Sport Science (University of Vigo, application 03-090425, 9 April 2025).

Data Availability Statement: The data presented in this study are openly available in FigShare at doi:10.6084/m9.figshare.29603429.

Acknowledgments: This publication was made possible thanks to the research stays during the years 2023 and 2024 at the Instituto Politécnico de Viana do Castelo [IPVC]—Escola Superior de Desporto e Lazer.

Conflicts of Interest: The authors declare no conflicts of interest.

References

1. Navia, J.A.; Ruiz, L.M. Análisis de la Complejidad Perceptivo-Motriz y Psicológica del Penalti en el Fútbol. *Rev. Int. Cienc. Deporte* **2014**, *10*, 264–280. [CrossRef]
2. Palao, J.M.; López-Montero, M.; López-Botella, M. Relación Entre Eficacia, Lateralidad y Zona de Lanzamiento del Penalti en Función del Nivel de Competición en Fútbol. (Relationship between Efficacy, Laterality of Foot Strike, and Shot Zone of the Penalty in Relation to Competition Level in Soccer). *Rev. Int. Cienc. Deporte* **2010**, *6*, 154–165. [CrossRef]
3. Lopes, J.E.; Araujo, D.; Peres, R.; Davids, K.; Barreiros, J. The Dynamics of Decision Making in Penalty Kick Situations in Association Football. *Open Sports Sci. J.* **2008**, *1*, 24–30. [CrossRef]
4. Barbero, J.R.; Lapresa Ajamil, D.; Arana, J.; Anguera, M.T. Sequential Analysis of the Interaction between Kicker and Goalkeeper in Penalty Kicks. *Cuad. Psicol. Deporte* **2024**, *24*, 208–224. [CrossRef]
5. Barbero, J.R.; Lapresa, D.; Arana, J.; Anguera, M.T. An Observational Analysis of Kicker–Goalkeeper Interaction in Penalties between National Football Teams in International Competitions. *Int. J. Perform. Anal. Sport* **2023**, *23*, 196–212. [CrossRef]

6. Almeida, C.; Volossovitch, A.; Duarte, R. Penalty Kick Outcomes in UEFA Club Competitions (2010–2015): The Roles of Situational, Individual and Performance Factors. *Int. J. Perform. Anal. Sport.* **2016**, *16*, 508–522. [CrossRef]
7. Hughes, M.; Wells, J. Analysis of Penalties Taken in Shoot-Outs. *Int. J. Perform. Anal. Sport.* **2002**, *2*, 55–72. [CrossRef]
8. Prieto-Lage, I.; Artigues-Ribas, L.; Gutiérrez-Santiago, A. Technical-Tactical Patterns of the Attacker and the Goalkeeper in the Penalty Shootouts within the Spanish Football League during the 2016-17 Season through T-Patterns and Polar Coordinates. *Cuad. Psicol. Deporte* **2019**, *20*, 166–180. [CrossRef]
9. Jamil, M.; Littman, P.; Beato, M. Investigating Inter-League and Inter-Nation Variations of Key Determinants for Penalty Success across European Football. *Int. J. Perform. Anal. Sport* **2020**, *20*, 892–907. [CrossRef]
10. Prieto-Lage, I.; Argibay-González, J.C.; Bezerra, P.; Cidre-Fuentes, P.; Reguera-López-de-la-Osa, X.; Gutiérrez-Santiago, A. Analysis of Penalty Kick Performance in the Spanish Football League: A Longitudinal Study. *Appl. Sci.* **2024**, *14*, 7046. [CrossRef]
11. Horn, M.; de Waal, S.; Kraak, W. In-Match Penalty Kick Analysis of the 2009/10 to 2018/19 English Premier League Competition. *Int. J. Perform. Anal. Sport* **2020**, *21*, 139–155. [CrossRef]
12. Almeida, C.; Volossovitch, A. Multifactorial Analysis of Football Penalty Kicks in the Portuguese First League: A Replication Study. *Int. J. Sports Sci. Coach.* **2022**, *18*, 160–175. [CrossRef]
13. Lopez-Botella, M.; Palao, J.-M. Relationship between Laterality of Foot Strike and Shot Zone on Penalty Efficacy in Specialist Penalty Takers. *Int. J. Perform. Anal. Sport* **2007**, *7*, 26–36. [CrossRef]
14. Bar-Eli, M.; Azar, O.H. Penalty Kicks in Soccer: An Empirical Analysis of Shooting Strategies and Goalkeepers' Preferences. *Soccer Soc.* **2009**, *10*, 183–191. [CrossRef]
15. Jordet, G.; Hartman, E.; Visscher, C.; Lemmink, K.A.P.M. Kicks from the Penalty Mark in Soccer: The Roles of Stress, Skill, and Fatigue for Kick Outcomes. *J. Sports Sci.* **2007**, *25*, 121–129. [CrossRef] [PubMed]
16. Memmert, D.; Hüttermann, S.; Hagemann, N.; Loffing, F.; Strauss, B. Dueling in the Penalty Box: Evidence-Based Recommendations on How Shooters and Goalkeepers Can Win Penalty Shootouts in Soccer. *Int. Rev. Sport. Exerc. Psychol.* **2013**, *6*, 209–229. [CrossRef]
17. Chiappori, P.-A.; Levitt, S.; Groseclose, T. Testing Mixed-Strategy Equilibria When Players Are Heterogeneous: The Case of Penalty Kicks in Soccer. *Am. Econ. Rev.* **2002**, *92*, 1138–1151. [CrossRef]
18. Furley, P.; Noël, B.; Memmert, D. Attention towards the Goalkeeper and Distraction during Penalty Shootouts in Association Football: A Retrospective Analysis of Penalty Shootouts from 1984 to 2012. *J. Sports Sci.* **2017**, *35*, 873–879. [CrossRef]
19. Pereira, M.R.; Patching, G.R. Goal Side Selection of Penalty Shots in Soccer: A Laboratory Study and Analyses of Men's World Cup Shoot-Outs. *Percept. Mot. Ski.* **2021**, *128*, 2279–2303. [CrossRef]
20. Morya, E.; Ranvaud, R.; Pinheiro, W.M. Dynamics of Visual Feedback in a Laboratory Simulation of a Penalty Kick. *J. Sports Sci.* **2003**, *21*, 87–95. [CrossRef]
21. Dalton, K.; Guillon, M.; Naroo, S.A. An Analysis of Penalty Kicks in Elite Football Post 1997. *Int. J. Sports Sci. Coach.* **2015**, *10*, 815–827. [CrossRef]
22. Csató, L. A Comparison of Penalty Shootout Designs in Soccer. *4OR* **2021**, *19*, 183–198. [CrossRef]
23. Furley, P.; Dicks, M.; Stendtke, F.; Memmert, D. "Get It out the Way. The Wait's Killing Me" Hastening and Hiding during Soccer Penalty Kicks. *Psychol. Sport. Exerc.* **2012**, *13*, 454–465. [CrossRef]
24. Anguera, M.T.; Blanco-Villaseñor, A.; Losada, J.L.; Portell, M. Guidelines for Designing and Conducting a Study That Applies Observational Methodology. *UB J. Psychol.* **2018**, *48*, 9–17. [CrossRef]
25. Soto-Fernández, A.; Camerino, O.; Iglesias, X.; Anguera, M.T.; Castañer, M. LINCE PLUS Software for Systematic Observational Studies in Sports and Health. *Behav. Res. Methods* **2022**, *54*, 1263–1271. [CrossRef]
26. Iván-Baragaño, I.; Ardá, A.; Anguera, M.T.; Losada, J.L.; Maneiro, R. Future Horizons in the Analysis of Technical-Tactical Performance in Women's Football: A Mixed Methods Approach to the Analysis of in-Depth Interviews with Professional Coaches and Players. *Front. Psychol.* **2023**, *14*, 1128549. [CrossRef]
27. Vázquez-Estévez, C.; Prieto-Lage, I.; Reguera-López-de-la-Osa, X.; Rodríguez-Crespo, M.; Gutiérrez-Santiago, J.A.; Gutiérrez-Santiago, A. Analysis and Successful Patterns in One-Possession Games During the Last Minute in the Women's EuroLeague. *Appl. Sci.* **2025**, *15*, 5046. [CrossRef]
28. Prieto-Lage, I.; Crespo, M.; Martínez-Gallego, R.; Reguera-López-de-la-Osa, X.; Silva-Pinto, A.J.; Gutiérrez-Santiago, A. First-Serve Advantage and Emerging Tactical Limitations in Elite U-14 Boys' Tennis: A Les Petits as Case Study. *Appl. Sci.* **2025**, *15*, 5341. [CrossRef]
29. Prieto-Lage, I.; Louzao-Neira, I.; Argibay-González, J.C.; Gutiérrez-Santiago, A. Injury Patterns of Professional Footballers in the Spanish First Division during the 2017–2018 Seasons. *Physiol. Behav.* **2020**, *224*, 113052. [CrossRef]

30. Magnusson, M.S.; Burgoon, J.K.; Casarrubea, M. *Discovering Hidden Temporal Patterns in Behavior and Interaction: T-Pattern Detection and Analysis with THEME*; Magnusson, M.S., Burgoon, J.K., Casarrubea, M., Eds.; Springer: New York, NY, USA, 2016; ISBN 978-1-4939-3248-1.
31. Amatria, M.; Lapresa, D.; Arana, J.; Anguera, M.; Jonsson, G. Detection and Selection of Behavioral Patterns Using Theme: A Concrete Example in Grassroots Soccer. *Sports* **2017**, *5*, 20. [CrossRef]
32. Eysenck, M.W.; Derakshan, N.; Santos, R.; Calvo, M.G. Anxiety and Cognitive Performance: Attentional Control Theory. *Emotion* **2007**, *7*, 336–353. [CrossRef] [PubMed]
33. Schmidt, R.A.; Lee, T.D. *Motor Learning and Performance: From Principles to Application*, 6th ed.; Human Kinetics: Champaign, IL, USA, 2019; ISBN 9781492574682.
34. Todorov, E.; Jordan, M.I. Optimal Feedback Control as a Theory of Motor Coordination. *Nat. Neurosci.* **2002**, *5*, 1226–1235. [CrossRef] [PubMed]
35. Davids, K.; Araujo, D.; Vilar, L. An Ecological Dynamics Approach to Skill Acquisition: Implications for Development of Talent in Sport. *Talent. Dev. Excell.* **2013**, *5*, 21–34.

symmetry

MDPI

Article

Asymmetry in the Alignment of School Furniture and Anthropometric Measures: A Comparative Study Between Two Schools in Spain and Portugal

Adrián Paramés-González [1], Alfonso Gutiérrez-Santiago [1,*], Agostinho Fernandes [2], Ana Cristina Braga [2], Nélson Costa [2,3], Paula Carneiro [2], Xoana Reguera-López-de-la-Osa [4,*] and Iván Prieto-Lage [1]

1 Observational Research Group, Faculty of Education and Sport, University of Vigo, 36005 Pontevedra, Spain; aparames@uvigo.es (A.P.-G.); ivanprieto@uvigo.es (I.P.-L.)

2 ALGORITMI Research Center/LASI, University of Minho, 4800-058 Guimarães, Portugal; agostinholfernandes@gmail.com (A.F.); acb@dps.uminho.pt (A.C.B.); ncosta@dps.uminho.pt (N.C.); pcarneiro@dps.uminho.pt (P.C.)

3 HTIR—Human-Technology Interaction and Robotics Lab, CCG ZGDV Institute, 4800-058 Guimarães, Portugal

4 Education, Physical Activity and Health Research Group (Gies10-DE3), Galicia Sur Health Research, Institute (IIS Galicia Sur), SERGAS-UVIGO, 36208 Vigo, Spain

* Correspondence: ags@uvigo.es (A.G.-S.); xreguera@uvigo.es (X.R.-L.-d.-l.-O.)

Abstract: Background: School ergonomics is a subject of growing interest for the scientific community due to the health problems that it is associated with in students, mainly asymmetries in the spine due to the use of chairs and desks that are inadequate for their anthropometry. This study aimed to analyze the anthropometric characteristics and asymmetries among fifth- to ninth-grade students in Spain and Portugal, with the goal of obtaining data on the ideal height of seats and desks. Additionally, it evaluated the correspondence in the recommended sizes of chairs and desks according to the parameters of the European Union catalog and examined the suitability of the height as a criterion for the allocation of school furniture. Methods: Different anthropometric variables, including the height, popliteal height, shoulder-to-seat height, and elbow-to-seat height, were measured in a stratified sample of 500 students (mean age = 12.7 years, SD = 1.2) across different grades (fifth grade = 86, sixth grade = 106, seventh grade = 95, eighth grade = 89, ninth grade = 124), genders (males = 256, females = 244), and countries (Spain = 191, Portugal = 309). These measurements were used to calculate the average ideal seat and desk heights based on anthropometric formulas, which were then compared to the current furniture allocation practices. The statistical analyses included *t*-tests, chi-squared tests, and effect sizes, with adjustments for multiple comparisons. Results: The results revealed significant asymmetries and low correspondence in the allocation of chairs and desks of the same sizes to students, with a match rate ranging between 40% and 70%. Moreover, the correspondence was even lower when using a formula based solely on height, compared to formulas validated with specific anthropometric measures, particularly for desks, where the asymmetries reached 100% in some grades. Conclusion: These findings highlight the need to improve the adaptation of school furniture to optimize student ergonomics and comfort, and they suggest disregarding the height as the primary criterion for furniture allocation. Additionally, assigning a desk size based on the recommended chair size is discouraged.

Keywords: education; anthropometry; ergonometry; seat; desk

1. Introduction

Posture and spinal symmetry are essential for physical development, especially during childhood and adolescence, when the body grows rapidly [1]. The spine, which serves as the central axis of the musculoskeletal system, plays a crucial role in the body's stability and mobility. Maintaining proper symmetry and posture is vital not only for musculoskeletal health but also for overall well-being [2]. During the early stages of life, postural habits are established that can prevent long-term back problems [3].

One of the most critical moments at which to focus on posture is in school. Students spend a significant amount of time sitting—approximately 6.5 h per day in school settings [4,5]. This prolonged sitting time, which can constitute between 70 and 90% of the school day [6,7], can lead to spinal asymmetries if proper posture is not maintained. In the short term, this may cause minor discomfort, but, over time, it can develop into more serious issues, such as scoliosis. In Spain, it has been found that 51% of boys and more than 69% of girls have experienced low back pain before the age of 15 [1,8]. Preventive intervention is crucial to prevent these problems from becoming chronic conditions in adulthood.

School furniture plays an essential role in the prevention of postural problems. A mismatch between the furniture dimensions and students' anthropometric characteristics can negatively impact their posture and long-term health [9–11].

Historically, the allocation of school furniture has been based on general criteria such as the chronological age or the average height of students. However, this approach does not take into account individual variations in body proportions or the differences in physical development among students of the same age. Therefore, it is essential to determine whether this method is suitable for the anthropometric characteristics of the student population.

The European Standard 1729:1-2015 provides guidelines on school furniture dimensions, defining criteria for the design of chairs and desks based on students' anthropometric characteristics [12]. This standard proposes eight different sizes to accommodate the variability in student dimensions throughout their education. To simplify allocation, it introduces a color-coding system that symmetrically associates each chair and desk size with a color. However, it is important to evaluate whether this symmetrical allocation truly matches the students' anthropometric characteristics. Previous research has shown that, in some educational centers, the implementation of this standard is insufficient, with the continued use of proprietary size guides rather than adhering to the European recommendations [13,14].

Early intervention is key to correcting postural misalignments before they become chronic problems. Postural education programs in schools, which teach students the importance of correct posture when sitting, walking, and engaging in daily activities, are essential [15]. In the classroom, the sitting posture is particularly relevant. Many students adopt incorrect postures, negatively affecting their spinal alignment. Proper posture involves keeping the back straight, shoulders relaxed, feet flat on the floor, and knees at a 90-degree angle, which evenly distributes the weight and reduces the spinal pressure [16], seeking adequate symmetry in the spine. School furniture should be adapted to the physical characteristics of students to avoid forced postures and promote a healthy learning environment [9,17]. Thus, the combination of ergonomic furniture design with proper postural education can reduce the incidence of musculoskeletal problems [18].

Early intervention with the effective implementation of ergonomic standards in school furniture and the correct adaptation of school furniture is crucial in preventing future musculoskeletal problems. Therefore, the objective of this study was to analyze the morphological asymmetries between two populations (Spain and Portugal) differentiated by sex, assess whether students in the same grade use the same size of school furniture, and verify whether the height is an adequate criterion for its allocation. Additionally, the correspondence between the chair and desk sizes of the European catalog based on color coding was examined.

To address the objectives of this study, the following research questions were formulated.

Are there significant differences in the anthropometric characteristics and furniture needs of students based on their grade, gender, and country of origin (Spain and Portugal)?

Does the alignment between school furniture sizes and students' anthropometric characteristics differ when using height-based criteria versus validated anthropometric formulas?

To what extent does the correspondence between chair and desk sizes meet the ergonomic needs of students as defined by the European school furniture catalog?

These questions aimed to explore the anthropometric disparities between the two populations and assess the adequacy of the current school furniture allocation practices.

2. Materials and Methods

2.1. Sample

The participants in this study were students from two public primary and secondary education schools located in a city in Southern Galicia and another in Northern Portugal. A total of 500 students (256 boys and 244 girls), with a mean age of 12.7 years (SD = 1.2), including boys (mean age = 12.10 years, SD = 1.44) and girls (mean age = 12.15 years, SD = 1.44), from grades 5 to 9 in Portugal and from the 5th and 6th grades of primary education and 1st to 3rd years of ESO (the equivalent of grades 7 to 9 in Portugal) in Spain, were invited to participate (see Table 1). A convenience sampling technique was used to facilitate the recruitment of the students. All students who provided signed informed consent from their parents or guardians, as well as their own assent, were included in the study. Repeating students were assigned to the grade corresponding to their biological age.

Table 1. Description of the study sample.

Grade	Spain (n = 191)		Portugal (n = 309)	
	Males (n = 91)	Females (n = 100)	Males (n = 165)	Females (n = 144)
5th grade	19	18	29	20
6th grade	25	23	25	33
7th grade	15	17	34	29
8th grade	13	19	36	21
9th grade	19	23	41	41
Total	91	100	165	144

Note: "Grade" refers to the academic year level of the students. For Spain, grades 5 and 6 correspond to primary education, while grades 7, 8, and 9 correspond to the 1st, 2nd, and 3rd years of ESO (secondary education). In Portugal, grades 5 to 9 represent consecutive years of basic education.

Students with diagnosed physical conditions or health issues that could affect their posture or anthropometric measurements (e.g., scoliosis, physical disabilities, or musculoskeletal disorders) were excluded from the study. This criterion ensured that the results reflected the typical anthropometric characteristics of the general student population.

To carry out this study, the necessary permissions were obtained from the school administration. All families and students were informed in advance about the research objectives, the procedures to be followed, the confidentiality statement, and the contact details of the researcher. The ethical principles of medical research involving human subjects, as outlined in the Declaration of Helsinki [19], were respected at all times.

The study received approval from the Ethics Committee of the Faculty of Education and Sport Sciences at the University of Vigo, under code 04/1019.

2.2. Measurements

The height, popliteal height, elbow height, and shoulder height were measured for this study. Anthropometric measurements followed the procedures established in previous, similar studies [9]. Students were assessed on the right side (except for height) while seated in an adjustable chair with a horizontal seat surface. The legs were kept at a 90° angle, with the feet fully resting on an adjustable footrest. During the measurement process, the students were barefoot and wore shorts and a short-sleeved shirt. A 60 cm Cescorf anthropometer, certified by the International Society for the Advancement of

Kinanthropometry (ISAK), was used for all measurements, with the exception of the height, which was recorded with a portable Seca stadiometer (range 20–205 cm). Measurements were conducted over ten sessions, one for each grade level evaluated, and recorded in centimeters by an assistant; all were carried out by the same anthropometrist. This approach minimized the errors that can occur when multiple anthropometrists are involved [20]. The precision and repeatability of the measurements were ensured by the anthropometrist's training, certified at ISAK Level 3, and their prior experience in such evaluations. At least two measurements were taken for each parameter, and, if the difference exceeded 0.5 cm, a third measurement was taken. The following anthropometric measures were included to calculate the ideal furniture dimensions [21] (see Figure 1).

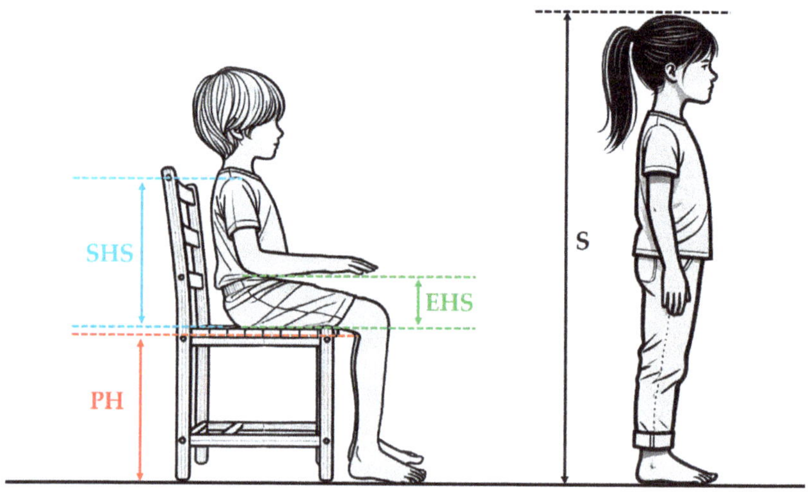

Figure 1. Anthropometric measurements.

- Stature (S): Defined as the vertical distance between the floor and the top of the head, measured with the subject standing upright and looking forward (Frankfurt plane).
- Seated shoulder height (SHS): The vertical distance from the sitting surface to the acromion.
- Seated elbow height (EHS): Measured with the elbow flexed at 90°. The vertical distance from the bottom of the elbow (olecranon) to the sitting surface.
- Popliteal height (PH): The knee should be bent at 90°. The vertical distance from the floor to the back of the knee (popliteal area).

The equations used to calculate the ideal heights of chairs and desks were as follows [22].

- Seat height (SH): $(PH+2.5)\cos30° \leq SH \leq (PH+2.5)\cos5°$;
- Desk height (DH): $(SH+EHS \leq DH \leq (SH + EHS*0.7396 + SHS*0.2604)$.

2.3. Procedure

The anthropometric measurements were carried out over five consecutive days, from Monday to Friday, in both countries. The assessments were conducted during school hours, between 9:00 a.m. and 2:00 p.m. Each day, one grade level was evaluated, starting with the 5th grade and ending with the 9th grade. In Spain, the measurements were taken during the second week of March 2023, while, in Portugal, they were conducted during the third week of the same month.

The data were recorded in an SPSS spreadsheet, where the ideal seat and desk heights for each subject were determined using the previously mentioned formulas. Additionally, the corresponding seat and desk size for each student was estimated based on their stature.

Subsequently, a correspondence analysis was performed between the size/color calculated from the anthropometric formulas, which included the popliteal height, elbow height, and shoulder height, and the estimation based solely on the stature. A discrepancy was considered when both calculations suggested a different size/color.

Furthermore, the degree of mismatch between the seat and desk size/color assigned to the students (see Figure 2) was evaluated using the formula based on multiple anthropometric measurements. A mismatch was considered when the size/color of one piece of furniture did not match the other, according to the parameters established in the reference furniture catalog of the European Union (e.g., yellow seat/size 3 and red desk/size 4).

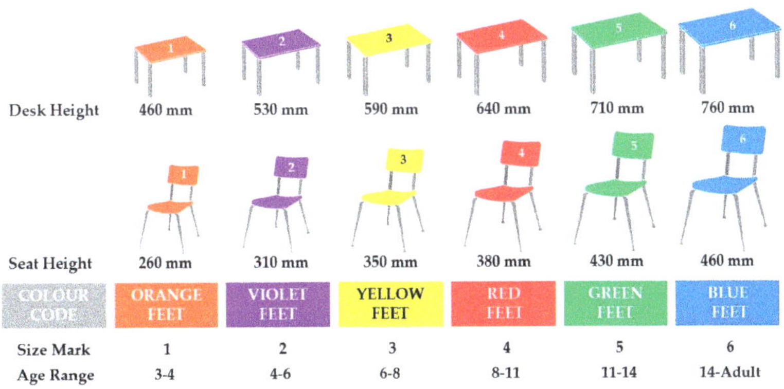

Figure 2. Seat and desk size types from the European Union's furniture catalog.

2.4. Data Analysis

All statistical analyses were performed using the IBM Statistical Package for the Social Sciences, version 25.0 (IBM SPSS Inc., Chicago, IL, USA). A descriptive analysis, stratified by grade level, was conducted for each of the study variables through measures of the central tendency (mean and standard deviation). The normality of the sample was checked using the Kolmogorov–Smirnov test (with Lilliefors correction) for variables with more than 50 cases and the Shapiro–Wilk test for variables with 50 or fewer cases. The mean values of the quantitative variables studied were compared between males and females and between Spain and Portugal, using a *t*-test for independent samples when the sample was normal and the Mann–Whitney U test when the sample was not normal. Qualitative variables, such as the grade, sex (male/female), country of origin (Spain/Portugal), and types of seat and desk sizes according to the EU catalog, were compared using crosstabs to calculate the chi-squared statistic (χ^2 test of independence) and evaluate associations between categorical data.

To control for the risk of committing a type I error due to multiple comparisons, the Holm–Bonferroni correction was applied within each set of comparisons, ensuring the rigorous adjustment of the significance levels. For comparisons that remained significant after applying the Holm–Bonferroni correction, effect sizes (Cohen's d) were calculated to assess the magnitude of the observed differences. The Cohen's d values were interpreted as small ($d = 0.2$), medium ($d = 0.5$), and large ($d = 0.8$).

In all statistical tests, a significance level of $p < 0.05$ was considered, except for those adjusted using the Holm–Bonferroni correction.

3. Results

Descriptive Analysis

Table 2 presents the descriptive analysis of the study, recording the measurements of the height, popliteal height, shoulder height while seated, and elbow height while seated with the elbow flexed at 90° to the seat. The analysis was performed by stratifying the data by country and by the students' academic grade.

Table 2. Descriptive analysis of the study.

Stature

Grade	Males N	Males $\overline{X}$	Males SD	Females N	Females $\overline{X}$	Females SD	t	p
Spain								
5th grade	19	142.3	9.2	18	145.2	7.1	−1.061	0.296
6th grade	25	148.3	8.4	23	149.6	5.7	−0.650	0.519
7th grade	15	151.7	7.4	17	151.2	6.1	0.233	0.817
8th grade	13	163.4	6.2	19	157.9	5.6	2.583	0.015
9th grade	19	169.9	7.4	23	157.7	6.5	5.697	0.000 *1
t/sex	1.119							
p	0.265							
Portugal								
5th grade	29	143.8	5.8	20	143.6	7.1	0.101	0.920
6th grade	25	146.1	4.9	33	148.1	7.1	−1.275	0.207
7th grade	34	156.3	9.4	29	156.4	8.4	−0.047	0.991
8th grade	36	161.4	7.2	21	159.4	6.3	1.049	0.299
9th grade	41	166.2	8.7	41	159.1	8.4	3.740	0.000
U o t/sex	10,639.500							
p	0.113							
U o t/Country	6899.5			−1.251				
p	0.284			0.212				

Popliteal Height

Grade	Males N	Males $\overline{X}$	Males SD	Females N	Females $\overline{X}$	Females SD	t	p
Spain								
5th grade	19	38.6	2.8	18	39.9	2.1	−1.657	0.106
6th grade	25	40.6	2.6	23	40.4	2.3	0.296	0.769
7th grade	15	39.8	2.2	17	38.6	1.5	1.761	0.091
8th grade	13	42.2	1.9	19	40.3	1.9	2.817	0.008 *2
9th grade	19	43.6	1.8	23	39.1	2.3	6.740	0.000 *3
t/sex	3.203							
p	0.002 *4							
Portugal								
5th grade	29	37.8	2.5	20	38.1	2.3	−0.404	0.688
6th grade	25	38.2	2.8	33	39.1	2.6	−1.216	0.229
7th grade	34	39.9	2.9	29	39.2	2.4	0.991	0.326
8th grade	36	40.7	2.8	21	40.1	2.8	0.804	0.425
9th grade	41	43.1	3.4	41	40.9	2.2	3.405	0.001
U o t/sex	1.729							
p	0.085							
U o t/Country	1.516			0.206				
p	0.131			0.837				

Elbow Height

Grade	Males N	Males $\overline{X}$	Males SD	Females N	Females $\overline{X}$	Females SD	t	p
Spain								
5th grade	19	18.5	2.2	18	19.0	1.8	−0.897	0.376
6th grade	25	19.6	2.4	23	19.9	2.1	−0.520	0.605
7th grade	15	17.8	2.1	17	20.0	2.5	−2.696	0.011
8th grade	13	19.0	2.0	19	20.0	1.7	−1.612	0.117
9th grade	19	19.5	2.3	23	20.6	2.8	−1.418	0.164
t/sex	−3.092							
p	0.002 *5							
Portugal								
5th grade	29	19.9	2.9	20	21.9	3.8	−1.954	0.059
6th grade	25	20.3	2.5	33	20.1	2.2	0.224	0.824
7th grade	34	22.5	3.7	29	24.4	3.9	−1.913	0.060
8th grade	36	22.8	3.7	21	24.7	4.2	−1.806	0.076
9th grade	41	23.6	3.7	41	25.5	3.6	−2.347	0.021
U o t/sex	9572.000							
p	0.003							
U o t/Country	3760.5			3471.0				
p	0.000 *6			0.000 *7				

Shoulder Height

Grade	Males N	Males $\overline{X}$	Males SD	Females N	Females $\overline{X}$	Females SD	t	p
Spain								
5th grade	19	47.3	3.9	18	49.3	2.4	−1.976	0.056
6th grade	25	49.6	3.4	23	50.7	2.6	−1.184	0.243
7th grade	15	49.5	2.7	17	51.2	3.2	−1.653	0.109
8th grade	13	53.8	2.9	19	52.5	2.4	1.310	0.200
9th grade	19	55.5	3.0	23	53.1	3.0	2.622	0.012
t/sex	−0.920							
p	0.359							
Portugal								
5th grade	29	50.1	4.3	20	51.6	4.6	−1.125	0.266
6th grade	25	50.0	3.6	33	51.3	3.8	−1.286	0.204
7th grade	34	55.4	5.3	29	56.4	5.3	−0.816	0.417
8th grade	36	56.5	3.6	21	56.7	3.9	−0.177	0.860
9th grade	41	58.4	4.8	41	58.7	4.3	−0.281	0.779
U o t/sex	−1.015							
p	0.311							
U o t/Country	−5.886			3907.0				
p	0.000 *8			0.000 *9				

Note: N = number of participants; $\overline{X}$ = mean; SD = standard deviation; t = t-statistic (result of t-test); U = Mann–Whitney U-statistic; p = level of significance. p-values were adjusted using the Holm–Bonferroni correction to control for multiple comparisons. Only values marked with an asterisk (*) remained significant after adjustment. Cohen's d values are provided for significant differences (Holm–Bonferroni corrected): [1] $d = 1.763$; [2] $d = 1.000$; [3] $d = 2.153$; [4] $d = 0.474$; [5] $d = -0.452$; [6] $d = -0.962$; [7] $d = -1.004$; [8] $d = -0.722$; [9] $d = -0.849$.

In the sample analyzed in Spain, statistically significant differences between males and females were observed in terms of the height and popliteal height in the eighth and ninth grades, as well as in the elbow-to-seat height in the seventh grade and the shoulder-to-seat height in the ninth grade. In all cases, males presented larger dimensions.

For the students in Portugal, significant differences between males and females were only found in the ninth grade for the variables of the height, popliteal height, and elbow-to-seat height. In all cases, except for the latter variable, males also showed larger dimensions.

No significant differences were found when comparing the results between Spain and Portugal in the variables of the height and elbow height. However, significant differences were observed in the elbow-to-seat height and shoulder-to-seat height, with larger dimensions in the Portuguese students in both cases.

Table 3 shows the mean ideal seat and desk height by age group for students from Spain and Portugal, together with a statistical comparison between the two countries.

Table 3. Descriptive analysis of ideal seat height and ideal desk height by grade and country.

Males	ISH Spain		ISH Portugal		*t*	*p*	IDH Spain		IDH Portugal		*t*	*p*
	$\overline{X}$	SD	$\overline{X}$	SD			$\overline{X}$	SD	$\overline{X}$	SD		
5th grade	38.2	2.6	37.5	2.3	0.997	0.324	60.4	4.2	61.3	3.0	−0.853	0.398
6th grade	40.1	2.4	37.9	2.6	3.205	0.002 *[1]	63.7	4.2	62.0	3.1	1.545	0.129
7th grade	39.4	2.1	39.5	2.7	−0.112	0.911	61.3	3.3	66.3	4.1	−4.151	0.000 *[2]
8th grade	41.7	1.7	40.3	2.6	1.789	0.080	65.2	2.6	67.4	3.4	−2.208	0.032 *[3]
9th grade	42.9	1.7	42.5	3.2	0.668	0.507	67.0	3.2	70.6	4.8	−2.917	0.005 *[4]
Females	**ISH Spain**		**ISH Portugal**		*t*	*p*	**IDH Spain**		**IDH Portugal**		*t*	*p*
	$\overline{X}$	SD	$\overline{X}$	SD			$\overline{X}$	SD	$\overline{X}$	SD		
5th grade	39.5	1.9	37.8	2.1	2.649	0.012 *[5]	62.5	2.7	63.5	4.8	−0.797	0.431
6th grade	39.9	2.1	38.7	2.4	2.004	0.050	63.9	3.9	62.9	3.5	1.013	0.315
7th grade	38.3	1.4	38.9	2.2	−1.055	0.297	62.3	3.3	67.4	4.2	−4.319	0.000 *[7]
8th grade	39.9	1.8	39.7	2.6	0.332	0.742	64.2	2.6	68.6	3.9	−4.117	0.000 *[8]
9th grade	38.8	2.2	40.5	2.0	−3.167	0.002 *[6]	63.6	2.9	70.3	4.1	−6.860	0.000 *[9]

Note: ISH = ideal seat height; IDH = ideal desk height; $\overline{X}$ = mean; SD = standard deviation; t = t-statistic (result of t-test); p = level of significance. p-values were adjusted using the Holm–Bonferroni correction to control for multiple comparisons; only values marked with an asterisk (*) remained significant after adjustment. Cohen's d values are provided for significant differences (Holm–Bonferroni corrected): [1] d = 0.881; [2] d = −1.289; [3] d = −0.684; [4] d = −0.820; [5] d = −0.847; [6] d = −0.820; [7] d = −1.309; [8] d = −1.314; [9] d = −1.802.

Significant differences were found in the ideal seat height for males in the fifth grade, as well as in the desk height in the seventh, eighth, and ninth grades, when comparing students from Spain and Portugal. For females, significant differences were also observed between students from both countries in the ideal seat height in the fifth, sixth, and ninth grades and in the desk height for the seventh to the ninth grades. In all cases, Portuguese students required higher furniture than Spanish students.

Table 4 provides a descriptive analysis of the seat sizes, according to the EU catalog, needed in classrooms from fifth to ninth grade, stratified by sex and country. Additionally, the analysis includes a comparison between students from Spain and Portugal.

The data showed that, considering the sex and country of origin, in the fifth grade, sizes 35 and 38 were predominant, with more than 72% of the students having anthropometric characteristics suitable for these two chair models. In the sixth grade, sizes 38 and 43 were the most common, with up to 68% of boys in Portugal requiring size 38. In the seventh grade, size 43 was the most frequent, with at least 55.2% of the students, regardless of sex and country, showing anthropometric dimensions compatible with this seat model. In the eighth and ninth grades, sizes 38 and 43 better suited the students' morphological characteristics compared to other sizes. It is worth noting that up to 84.2% of Spanish boys needed size 43, while up to 60.9% of Spanish girls required size 38. It was found that using two sizes per grade, based on the needs, allowed the coverage of over 70% of the students, regardless of their sex and country of origin.

Table 4. Types of seat sizes according to EU catalog by grade and country.

Grade		Males						Females					
		Spain		Portugal		χ^2	p	Spain		Portugal		χ^2	p
		n	%	n	%				%	n	%		
5th grade	S26	0	0.0%	0	0.0%	3.586	0.310	0	0.0%	0	0.0%	3.022	0.388
	S31	1	5.3%	1	3.4%			0	0.0%	1	5.0%		
	S35	2	10.5%	10	34.5%			2	11.1%	4	20.0%		
	S38	13	68.4%	14	48.3%			11	61.1%	13	65.0%		
	S43	3	15.8%	4	13.8%			5	27.8%	2	10.0%		
	S46	0	0.0%	0	0.0%			0	0.0%	0	0.0%		
6th grade	S26	0	0.0%	0	0.0%	11.019	0.012		0.0%	0	0.0%	3.067	0.216
	S31	0	0.0%	4	16.0%			0	0.0%	0	0.0%		
	S35	2	8.0%	1	4.0%			1	4.3%	6	18.2%		
	S38	11	44.0%	17	68.0%			13	56.5%	19	57.6%		
	S43	12	48.0%	3	12.0%			9	39.1%	8	24.2%		
	S46	0	0.0%	0	0.0%			0	0.0%	0	0.0%		
7th grade	S26	0	0.0%	0	0.0%	1.37	0.713		0.0%	0	0.0%	5.336	0.069
	S31	0	0.0%	0	0.0%			0	0.0%	0	0.0%		
	S35	1	6.7%	5	14.7%			1	5.9%	7	24.1%		
	S38	9	60.0%	20	58.8%			15	88.2%	16	55.2%		
	S43	5	33.3%	8	23.5%			1	5.9%	6	20.7%		
	S46	0	0.0%	1	2.9%			0	0.0%	0	0.0%		
8th grade	S26	0	0.0%	0	0.0%	2.123	0.547		0.0%	0	0.0%	1.905	0.592
	S31	0	0.0%	0	0.0%			0	0.0%	0	0.0%		
	S35	0	0.0%	2	5.6%			0	0.0%	1	4.8%		
	S38	3	23.1%	14	38.9%			12	63.2%	12	57.1%		
	S43	9	69.2%	18	50.0%			7	36.8%	7	33.3%		
	S46	1	7.7%	2	5.6%			0	0.0%	1	4.8%		
9th grade	S26	0	0.0%	0	0.0%	6.868	0.076		0.0%	0	0.0%	10.688	0.005
	S31	0	0.0%	0	0.0%			0	0.0%	0	0.0%		
	S35	0	0.0%	1	2.4%			4	17.4%	0	0.0%		
	S38	2	10.5%	13	31.7%			14	60.9%	20	48.8%		
	S43	16	84.2%	20	48.8%			5	21.7%	21	51.2%		
	S46	1	5.3%	7	17.1%			0	0.0%	0	0.0%		

Note: n = number of participants; % = percentage; χ^2 = chi-squared statistic; p = level of significance; S26 = seat size 26; S31 = seat size 31; S35 = seat size 35; S38 = seat size 38; S43 = seat size 43; S46 = seat size 46.

The analysis indicated that, except for sixth-grade boys and ninth-grade girls, there were no significant differences in the required chair sizes for classroom furniture from the fifth to ninth grade between students from Spain and Portugal.

The significant differences mentioned earlier indicated that, in the sixth grade in Spain, more boys required taller seats compared to Portugal. Specifically, more boys in Spain needed size 43 (48%) than in Portugal (12%). In the ninth grade, the opposite occurred: in Portugal, more girls needed size 44 compared to Spain (51.2% versus 21.7%).

In Table 5, the same results are presented as in the previous table but regarding desks.

The analysis showed a predominance of sizes 59 and 64 for both sexes and countries in fifth grade. It is important to note that 55.2% of Portuguese boys and 72.2% of Spanish girls required size 64. In sixth grade, most students also needed size 64 (ranging from 43.5% to 64%), although a considerable percentage of boys and girls required size 59 (ranging between 16% and 33.3%, depending on sex and country). In seventh grade, the predominant sizes were 64 and 71, with a minimum of 34.5% of Portuguese girls and a maximum of 52.9% of Spanish girls needing size 64. Notably, 51.7% of Portuguese girls required size 71. In eighth grade, size 64 was clearly dominant, being needed by up to 76.9% of Spanish boys, 52.8% of Portuguese boys, 73.7% of Spanish girls, and 42.9% of

Portuguese girls. Finally, in ninth grade, sizes 64 and 71 accommodated the majority of students, with more than half of Spanish boys and 53.7% of Portuguese girls requiring size 71, although only 13% of Spanish girls needed this size. It is worth noting that, in Portugal, a large proportion of students also needed size 76 (22.4% of boys and 22.0% of girls). Again, as with the chairs, it was found that using only two sizes could accommodate, in the worst-case scenario, up to 70% of students.

Table 5. Types of desk sizes according to EU catalog by grade and country.

Grade		Males						Females					
		Spain		Portugal		χ^2	p	Spain		Portugal		χ^2	p
		n	%	n	%			n	%	n	%		
5th grade	D46	0	0.0%	0	0.0%	4.431	0.219		0.0%	0	0.0%	6.436	0.092
	D53	3	15.8%	1	3.4%			0	0.0%	1	5.0%		
	D59	8	42.1%	12	41.4%			5	27.8%	7	35.0%		
	D64	7	36.8%	16	55.2%			13	72.2%	8	40.0%		
	D71	1	5.3%	0	0.0%			0	0.0%	4	20.0%		
	D76	0	0.0%	0	0.0%			0	0.0%	0	0.0%		
6th grade	D46	0	0.0%	0	0.0%	4.366	0.359		0.0%	0	0.0%	5.058	0.168
	D53	2	8.0%	2	8.0%			0	0.0%	1	3.0%		
	D59	4	16.0%	8	32.0%			7	30.4%	11	33.3%		
	D64	16	64.0%	15	60.0%			10	43.5%	19	57.6%		
	D71	2	8.0%	0	0.0%			6	26.1%	2	6.1%		
	D76	1	4.0%	0	0.0%			0	0.0%	0	0.0%		
7th grade	D46	0	0.0%	0	0.0%	6.968	0.073		0.0%	0	0.0%	14.969	0.002
	D53	0	0.0%	0	0.0%			0	0.0%	0	0.0%		
	D59	8	53.3%	7	20.6%			7	41.2%	2	6.9%		
	D64	6	40.0%	15	44.1%			9	52.9%	10	34.5%		
	D71	1	6.7%	10	29.4%			1	5.9%	15	51.7%		
	D76	0	0.0%	2	5.9%			0	0.0%	2	6.9%		
8th grade	D46	0	0.0%	0	0.0%	3.702	0.157		0.0%	0	0.0%	8.778	0.032
	D53	0	0.0%	0	0.0%			0	0.0%	0	0.0%		
	D59	1	7.7%	1	2.8%			2	10.5%	0	0.0%		
	D64	10	76.9%	19	52.8%			14	73.7%	9	42.9%		
	D71	2	15.4%	16	44.4%			3	15.8%	10	47.6%		
	D76	0	0.0%	0	0.0%			0	0.0%	2	9.5%		
9th grade	D46	0	0.0%	0	0.0%	11.154	0.011		0.0%	0	0.0%	26.471	0.000
	D53	0	0.0%	0	0.0%			0	0.0%	0	0.0%		
	D59	0	0.0%	2	4.9%			5	21.7%	0	0.0%		
	D64	9	47.4%	6	14.6%			15	65.2%	10	24.4%		
	D71	10	52.6%	23	56.1%			3	13.0%	22	53.7%		
	D76	0	0.0%	10	24.4%			0	0.0%	9	22.0%		

Note: n = number of participants; % = percentage; χ^2 = chi-squared statistic; p = level of significance; D46 = desk size 46; D53 = desk size 53; D59 = desk size 59; D64 = desk size 64; D71 = desk size 71; D76 = desk size 76.

For boys, no statistically significant differences were found when comparing students from both countries in terms of the size required in class, except in ninth grade. At this level, a large percentage of Portuguese boys required size 76 (56.1%), which did not occur among Spanish students (0%).

Regarding girls, there were no differences in the fifth and sixth grades, but there were differences in the seventh to ninth grades. In these cases, Portuguese girls generally required larger sizes in a greater percentage than Spanish students.

The following presents an analysis of the asymmetry in the matching of school furniture to students. First, as shown in Figure 3, a general evaluation was conducted, without stratification by sex, grade, or country, to determine whether the chair size, as indicated by the color in the European Union catalog, matched the desk size of the corresponding color (e.g., chair size 26 cm —red— with desk size 43 cm —red—). Additionally, the agreement was compared between the recommended size for a student, obtained through formulas based on specific anthropometric measurements (popliteal height, elbow-to-seat distance, and shoulder-to-seat distance), and the size determined solely based on the student's height.

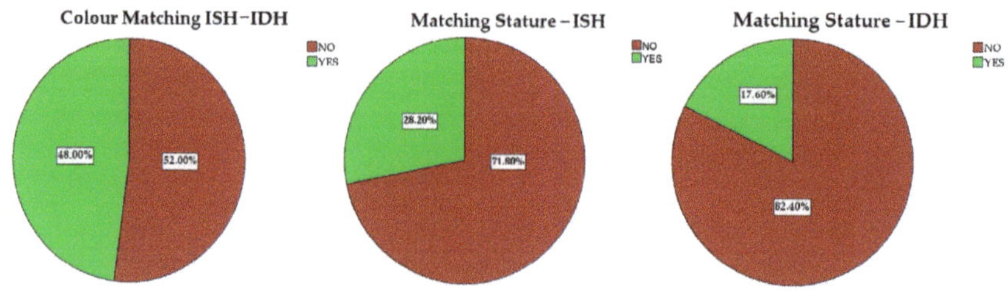

Figure 3. Color-coded matching between ideal sitting height (ISH) and ideal desk height (IDH), matching stature to ISH, and matching stature to IDH (global data, not stratified by sex, country, or grade level).

The results showed a mismatch of over 50% when comparing the alignment between seat and desk sizes of the same color according to the European Union catalog. This percentage increased to more than 70% when the student's height was used as the sole criterion to assign furniture.

Table 6 presents the previously described asymmetry analysis, but taking into account the students' sex, grade level, and country.

Table 6. Degree of concordance of seat and desk size according to the European Union catalog (based on pre-set color) and degree of agreement in the calculation of the ideal seat and desk height according to the height formula or the validated anthropometric measures formula.

Grade			Males						Females					
			Spain		Portugal		χ^2	p	Spain		Portugal		χ^2	p
			n	%	n	%			n	%	n	%		
5th grade	Color Matching ISH–IDH	No	13	68.4%	14	48.3%	1.893	0.169	11	61.1%	13	65.0%	0.062	0.804
		Yes	6	31.6%	15	51.7%			7	38.9%	7	35.0%		
	Matching Stature–ISH	No	9	47.4%	15	51.7%	0.087	0.768	8	44.4%	11	55.0%	0.422	0.516
		Yes	10	52.6%	14	48.3%			10	55.6%	9	45.0%		
	Matching Stature–IDH	No	14	73.7%	23	79.3%	0.206	0.650	11	61.1%	16	80.0%	1.643	0.200
		Yes	5	26.3%	6	20.7%			7	38.9%	4	20.0%		
6th grade	Color Matching ISH–IDH	No	15	60.0%	12	48.0%	0.725	0.395	12	52.2%	13	39.4%	0.896	0.344
		Yes	10	40.0%	13	52.0%			11	47.8%	20	60.6%		
	Matching Stature–ISH	No	12	48.0%	16	64.0%	1.299	0.254	11	47.8%	21	63.6%	1.383	0.240
		Yes	13	52.0%	9	36.0%			12	52.2%	12	36.4%		
	Matching Stature–IDH	No	19	76.0%	22	88.0%	1.22	0.269	16	69.6%	31	93.9%	5.97	0.015
		Yes	6	24.0%	3	12.0%			7	30.4%	2	6.1%		
7th grade	Color Matching ISH–IDH	No	11	73.3%	18	52.9%	1.792	0.181	6	35.3%	19	65.5%	3.946	0.047
		Yes	4	26.7%	16	47.1%			11	64.7%	10	34.5%		
	Matching Stature–ISH	No	9	60.0%	26	76.5%	1.384	0.240	15	88.2%	24	82.8%	0.249	0.618
		Yes	6	40.0%	8	23.5%			2	11.8%	5	17.2%		
	Matching Stature–IDH	No	15	100.0%	28	82.4%	3.016	0.082	17	100.0%	19	65.5%	7.49	0.006
		Yes	0	0.0%	6	17.6%			0	0.0%	10	34.5%		
8th grade	Color Matching ISH–IDH	No	10	76.9%	17	47.2%	3.406	0.065	6	31.6%	14	66.7%	4.912	0.027
		Yes	3	23.1%	19	52.8%			13	68.4%	7	33.3%		
	Matching Stature–ISH	No	12	92.3%	29	80.6%	0.966	0.326	18	94.7%	16	76.2%	2.691	0.101
		Yes	1	7.7%	7	19.4%			1	5.3%	5	23.8%		
	Matching Stature–IDH	No	13	100.0%	36	100.0%			19	100.0%	16	76.2%	5.17	0.023
		Yes	0	0.0%	0	0.0%			0	0.0%	5	23.8%		

Table 6. *Cont.*

Grade			Males						Females					
			Spain		Portugal		χ^2	p	Spain		Portugal		χ^2	p
			n	%	n	%			n	%	n	%		
9th grade	Color Matching ISH–IDH	No	8	42.1%	18	43.9%	0.017	0.896	9	39.1%	21	51.2%	0.865	0.352
		Yes	11	57.9%	23	56.1%			14	60.9%	20	48.8%		
	Matching Stature–ISH	No	18	94.7%	33	80.5%	2.068	0.150	22	95.7%	34	82.9%	2.181	0.140
		Yes	1	5.3%	8	19.5%			1	4.3%	7	17.1%		
	Matching Stature–IDH	No	19	100.0%	29	70.7%	6.951	0.008	23	100.0%	26	63.4%	10.991	0.001
		Yes	0	0.0%	12	29.3%			0	0.0%	15	36.6%		

Note: n = number of participants; % = percentage; χ^2 = chi-squared statistic; p = level of significance; ISH = ideal seat height; IDH = ideal desk height.

The highest alignment was found in Spanish girls in the eighth grade, with 68.4%. In contrast, the greatest asymmetry was also found in the eighth grade but among Spanish boys, with 76.9%. Generally, the asymmetry ranged between 40% and 70%.

When the alignment was evaluated by determining the chair size using only the height, compared to the validated formula that incorporated several anthropometric measurements, a high degree of discordance was found. The best alignment was recorded in the fifth grade, with 52.6% for Spanish boys, but many alignments fell below 20%, especially in higher grades. In the fifth and sixth grades, the alignment ranged from 40% to 60% but then decreased significantly.

For the desks, the asymmetry was even greater, with very low alignment percentages from the fifth to the ninth grades. The best alignment, considering the grade, sex, and country, was 38.9% in Spanish girls in the fifth grade. In several grades in both countries, there was 100% discordance.

4. Discussion

The aim of this study was to analyze the morphological asymmetries between two populations differentiated by grade, gender, and nationality (Spain and Portugal). Additionally, it evaluated whether students in the same grade used a uniform size of school furniture. Furthermore, the study aimed to verify the degree of correspondence between the assignment of a specific size and color of chair and the corresponding type of desk, according to the European Union's school furniture catalog. Finally, it examined whether the height was an appropriate criterion for the assignment of school furniture

This approach is highly relevant since inadequate school furniture is associated with a higher prevalence of musculoskeletal problems, especially at early ages, representing critical stages for posture and growth [23]. Moreover, morphological asymmetries and their relationships with furniture not only affect students' comfort and efficiency but also directly influence their long-term health [24]. In this regard, this study provides key data to better understand the need for a more precise and personalized approach to school furniture design.

This study's results confirm the trends observed in previous research [17,25]. In the fifth and sixth grades, girls tend to be taller than boys. However, in the seventh grade, the height differences between the sexes almost disappear. From the eighth grade onwards, a notable difference reappears, but in favor of boys, who show a tendency to be taller than their female classmates.

This growth pattern is related to the phases of pubertal development, which tend to occur earlier in girls. Anthropometric studies indicate that girls usually reach their growth peak before boys, explaining the difference observed in the early years of secondary education. However, once they reach puberty, boys experience a more prolonged and accelerated period of growth, eventually surpassing girls in height by the end of adolescence [26,27].

This phenomenon highlights the importance of considering not only sex but also the stage of biological development when designing and assigning school furniture [14,28].

National differences can also be related to genetic, nutritional, and socioeconomic factors, although these are not extremely significant between Spain and Portugal. Despite the cultural and geographical similarities between both countries, some studies have suggested that small variations in diet, physical activity, and access to healthcare may influence students' growth characteristics [29]. However, the differences observed in this study are not pronounced enough to justify country-specific school furniture designs, reaffirming the viability of a common approach in Europe [12], as long as the diversity within each school population is considered.

The data on ideal chair and desk heights clearly show that it is not easy to standardize school furniture sizes by academic grade. There is no grade, not even when separating students by sex in Spain and Portugal, in which a single size of chair or desk can be suitable for all students.

This finding confirms what previous studies have indicated [14,28,30]: the variability in anthropometric dimensions within each school group is too large for a uniform approach to work effectively. Standardization based on the academic grade ignores the diversity in students' growth and morphologies, potentially leading to a significant mismatch between ergonomic needs and the available furniture. Therefore, it is essential that school furniture design and assignment consider greater adaptability, allowing for personalized adjustments that accommodate individual variations.

Even when separating students by sex, the results indicate that at least two different sizes of chairs and desks are needed per grade. In the most complex scenarios, up to four different sizes of chairs and desks were required within the same grade and sex [25], highlighting the need for more adaptable and personalized furniture.

The wide variability in furniture sizes within the same grade emphasizes the importance of offering a broader range of furniture options, possibly through adjustable solutions [14,31–33]. The use of height-adjustable chairs and desks would allow for more precise adaptation to each student's needs, promoting proper posture and reducing the risk of long-term musculoskeletal problems. This finding supports the need for educational policies that promote investment in flexible and ergonomically efficient furniture.

On the other hand, as previous research has shown [25,34], this study confirms that the strategy of assigning school furniture based solely on height is inadequate. There is considerable disagreement between the estimated chair and desk heights when using this technique and the students' actual needs when validated formulas considering multiple anthropometric measurements are used.

This discrepancy highlights that the height is not a sufficient measure to determine a student's ergonomic needs. Body proportions, such as the torso length, the arm length, and the distance from the floor to the thighs, are equally important in ensuring proper posture while seated [34]. Validated anthropometric formulas, which include these additional measurements, offer a more precise approach, but their widespread application in the school setting requires training and tools that are not currently available in all educational institutions [35].

Although not all teachers have the necessary training to use an anthropometer correctly, which could lead to inaccurate measurements, recent proposals have emerged that allow furniture size estimation without relying exclusively on students' height.

These proposals include the use of digital applications and automated measurement tools, which could facilitate more accurate and efficient furniture assignment [17]. Some of these technologies can be easily integrated into the school environment, reducing the burden on teachers and improving the measurement accuracy, allowing for the greater individualization of the assigned furniture.

Moreover, this study also demonstrates that assigning desks based on the chair size, solely determined by popliteal height measurements, is not an effective strategy. The data revealed that the number of students whose chair size matched the corresponding desk size was alarmingly low. The popliteal height, while useful in determining the chair size, does not adequately consider the relationship between the leg and torso lengths, which is essential in determining the optimal desk height. This finding is consistent with previous research [32] that suggests that basing both decisions on a single anthropometric measurement is insufficient and potentially harmful to students' posture. The European UNE-EN 1729 standard for school furniture, which associates chairs and desks by size and color, suggests a direct relationship between the two, but this study's results indicate that this correlation is deficient.

This study questions the viability of the current European standard for school furniture, which rigidly associates chairs and desks. The findings suggest that a review of these standards is necessary to incorporate greater flexibility in furniture assignment, possibly adopting adjustable systems that can be modified according to individual students' needs.

4.1. Practical Implications

The results of this study have relevant implications for the design and allocation of school furniture in the European context. Firstly, it is clear that the variability in anthropometric dimensions within the same school grade, both in Spain and Portugal, requires a more personalized approach in the selection of furniture. Adaptable furniture with various size options for chairs and desks is necessary to allow students to maintain an ergonomically correct posture, thus reducing the risk of postural and musculoskeletal problems.

Additionally, educational institutions and school furniture manufacturers should reconsider the strategy of assigning furniture based solely on the height or chair size, as these methods do not ensure precise correspondence between chair and desk sizes. Implementing more comprehensive ergonomic assessment tools, which include various anthropometric measurements, could significantly improve the suitability of the furniture and its impact on students' postural health. It would also be beneficial to introduce training programs for teachers and administrative staff on ergonomic evaluation and the proper use of anthropometric measurement instruments.

4.2. Limitations and Future Perspectives

This study, while rigorous, presents some limitations that must be considered when interpreting the results. Firstly, the analysis focused on a sample of students from two specific schools, one in Southern Galicia (Spain) and another in Northern Portugal. While the findings provide valuable insights into the anthropometric characteristics and furniture alignment of students in these locations, they may not be generalizable to the entire populations of Spain and Portugal or to other European populations. Cultural, socioeconomic, and genetic differences, as well as variability in school furniture standards in other regions, could influence the results. It would be advisable to replicate the study in a broader and more diverse sample, including a multicenter approach to validate and expand upon these findings.

Another limitation is the reliance on anthropometric measurements obtained at a single point in time, which may not reflect changes in body dimensions throughout the school year. Growth patterns, especially during adolescence, can vary over short periods, affecting the validity of furniture recommendations based on these specific measurements. Furthermore, other factors, such as the posture adopted by students while using the furniture, were not analyzed, but these could impact their comfort and long-term health.

This study opens the door for several future research lines. Firstly, it would be interesting to expand the analysis at the European level, comparing the morphological asymmetries and the adequacy of school furniture in a broader context that includes countries with different socioeconomic and cultural backgrounds. This would allow for the evaluation of the effectiveness of the European school furniture catalog in different scenarios and establish recommendations that are more tailored to the reality of each country.

Additionally, longitudinal studies could be conducted to follow students over several years to observe how their body dimensions and furniture needs vary during their development. It would also be useful to investigate the relationship between the adequacy of school furniture and academic outcomes, as well as its impact on the prevention of health issues, such as back pain or postural asymmetries.

Finally, future research could focus on developing new tools and methodologies for the ergonomic evaluation of school furniture that are accessible to both educational staff and manufacturers, aiming to improve the accuracy in furniture allocation and optimize the educational experience for students.

5. Conclusions

This study demonstrates that, due to the significant variability in the anthropometric dimensions of students, it is not possible to standardize school furniture by grade level in Spain and Portugal. Even separating students by sex does not make it feasible to apply a single size for chairs or desks, highlighting the need for more adaptable and personalized furniture for each grade.

Furthermore, the system that assigns furniture solely based on height, or that links the size of the desk to that of the chair, proves ineffective. The mismatch between these two pieces of furniture indicates that the European catalog of chairs and desks by size does not meet the real needs of students.

Therefore, it is essential to adopt a more comprehensive approach that considers various anthropometric measurements to ensure an appropriate ergonomic posture. This will not only enhance the educational experience but also help to prevent long-term postural and health problems.

Finally, this study emphasizes the importance of continuing such research in other European populations and developing more accessible tools that allow for the estimation of suitable furniture without relying exclusively on the student's height.

Author Contributions: Conceptualization, A.P.-G., A.G.-S., A.F., P.C. and I.P.-L.; methodology, A.P.-G., A.G.-S., A.F., A.C.B., N.C. and I.P.-L.; software, A.G.-S., X.R.-L.-d.-l.-O. and I.P.-L.; validation, A.C.B., X.R.-L.-d.-l.-O. and I.P.-L.; formal analysis, A.G.-S., A.C.B., P.C., X.R.-L.-d.-l.-O. and I.P.-L.; investigation, A.P.-G., A.G.-S., A.F., A.C.B., N.C., P.C., X.R.-L.-d.-l.-O. and I.P.-L.; resources, A.P.-G., A.G.-S., A.F., N.C., P.C., X.R.-L.-d.-l.-O. and I.P.-L.; data curation, A.P.-G., A.F., A.C.B., N.C., P.C., X.R.-L.-d.-l.-O. and I.P.-L.; writing—original draft, A.P.-G., A.G.-S., P.C., X.R.-L.-d.-l.-O. and I.P.-L.; writing—review and editing, A.P.-G., A.G.-S., A.C.B., N.C., P.C. and I.P.-L.; visualization, A.F., N.C. and X.R.-L.-d.-l.-O.; supervision, A.P.-G., A.G.-S., A.C.B., N.C., P.C., X.R.-L.-d.-l.-O. and I.P.-L.; project administration, A.P.-G., A.G.-S., A.C.B., N.C., X.R.-L.-d.-l.-O. and I.P.-L.; funding acquisition, I.P.-L. and A.G.-S. All authors have read and agreed to the published version of the manuscript.

Funding: This research received no external funding.

Institutional Review Board Statement: This study was approved by the ethics committee of the Faculty of Education and Sport Science (University of Vigo, application 08-280722).

Informed Consent Statement: Informed consent was obtained from all subjects involved in the study. The parents or legal guardians of the participating students provided signed consent, and the students provided their assent.

Data Availability Statement: The raw data supporting the conclusions of this article will be made available by the authors on request.

Conflicts of Interest: The authors declare no conflicts of interest.

References

1. Calvo-Muñoz, I.; Kovacs, F.M.; Roqué, M.; Fernández, I.G.; Calvo, J.S. Risk Factors for Low Back Pain in Childhood and Adolescence: A Systematic Review. *Clin. J. Pain* **2018**, *34*, 468–484. [CrossRef] [PubMed]
2. Sharma, S.; Rawat, V. The Importance of Body Posture in Adolescence and its Relationship with Overall Well-being. *Indian J. Med. Spec.* **2023**, *14*, 197–205. [CrossRef]

3. Araújo, C.L.; Moreira, A.; Carvalho, G.S. Postural Education Programmes with School Children: A Scoping Review. *Sustainability* **2023**, *15*, 10422. [CrossRef]
4. Cooper, A.R.; Goodman, A.; Page, A.S.; Sherar, L.B.; Esliger, D.W.; van Sluijs, E.M.F.; Andersen, L.B.; Anderssen, S.; Cardon, G.; Davey, R.; et al. Objectively measured physical activity and sedentary time in youth: The International children's accelerometry database (ICAD). *Int. J. Behav. Nutr. Phys. Act.* **2015**, *12*, 113. [CrossRef]
5. Janssen, X.; Mann, K.D.; Basterfield, L.; Parkinson, K.N.; Pearce, M.S.; Reilly, J.K.; Adamson, A.J.; Reilly, J.J. Development of sedentary behavior across childhood and adolescence: Longitudinal analysis of the Gateshead Millennium Study. *Int. J. Behav. Nutr. Phys. Act.* **2016**, *13*, 88. [CrossRef]
6. Mooses, K.; Mägi, K.; Riso, E.M.; Kalma, M.; Kaasik, P.; Kull, M. Objectively measured sedentary behaviour and moderate and vigorous physical activity in different school subjects: A cross-sectional study. *BMC Public Health* **2017**, *17*, 108. [CrossRef]
7. Schwenke, P.; Coenen, M. Influence of Sit-Stand Tables in Classrooms on Children's Sedentary Behavior and Teacher's Acceptance and Feasibility: A Mixed-Methods Study. *Int. J. Environ. Res. Public Health* **2022**, *19*, 6727. [CrossRef]
8. Miñana-Signes, V.; Monfort-Pañego, M.; Bosh-Bivià, A.H.; Noll, M. Prevalence of Low Back Pain among Primary School Students from the City of Valencia (Spain). *Healthcare* **2021**, *9*, 270. [CrossRef]
9. Castellucci, H.I.; Arezes, P.M.; Molenbroek, J.F.M.; de Bruin, R.; Viviani, C. The influence of school furniture on students' performance and physical responses: Results of a systematic review. *Ergonomics* **2016**, *60*, 93–110. [CrossRef]
10. Arefi, M.F.; Pouya, A.B.; Poursadeqiyan, M. Investigating the match between anthropometric measures and the classroom furniture dimensions in Iranian students with health approach. *J. Educ. Health Promot.* **2021**, *10*, 38. [CrossRef]
11. Kahya, E. Mismatch between classroom furniture and anthropometric measures of university students. *Int. J. Ind. Ergon.* **2019**, *74*, 102864. [CrossRef]
12. CEN. *Furniture—Chairs and Tables for Educational Institutions—Part 1: Functional Dimensions*; CEN (European Committee for Standadization): Brussels, Belgium; European Union: Brussels, Belgium, 2015.
13. Macedo, A.C.; Morais, A.V.; Martins, H.F.; Martins, J.C.; Pais, S.M.; Mayan, O.S. Match between classroom dimensions and students' anthropometry: Re-equipment according to european educational furniture standard. *Hum. Factors* **2015**, *57*, 48–60. [CrossRef] [PubMed]
14. Paramés-González, A.; Gutiérrez-Santiago, A.; Gutiérrez-Santiago, J.A.; Prieto-Lage, I. La prevención del dolor de espalda mediante la correcta asignación del mobiliario escolar: Validación de dos instrumentos. *Rev. Esp. Salud Publica* **2021**, *95*, e1–e15.
15. Vidal, J.; Borras, P.A.; Ortega, F.B.; Cantallops, J.; Ponseti, X.; Palou, P. Effects of postural education on daily habits in children. *Int. J. Sports Med.* **2011**, *32*, 303–308. [CrossRef]
16. Cho, M.; Han, J.-S.; Kang, S.; Ahn, C.-H.; Kim, D.-H.; Kim, C.-H.; Kim, K.-T.; Kim, A.-R.; Hwang, J.-M. Biomechanical Effects of Different Sitting Postures and Physiologic Movements on the Lumbar Spine: A Finite Element Study. *Bioengineering* **2023**, *10*, 1051. [CrossRef]
17. Gutiérrez-Santiago, A.; Prieto-Lage, I.; Cancela-Carral, J.M.; Paramés-González, A. Validation of Two Instruments for the Correct Allocation of School Furniture in Secondary Schools to Prevent Back Pain. *Int. J. Environ. Res. Public Health* **2021**, *19*, 20. [CrossRef] [PubMed]
18. Ansari, S.; Nikpay, A.; Varmazyar, S. Design and Development of an Ergonomic Chair for Students in Educational Settings. *Health Scope* **2018**, *in press*. [CrossRef]
19. Harriss, D.J.; Atkinson, G. Ethical Standards in Sport and Exercise Science Research: 2016 Update. *Int. J. Sports Med.* **2015**, *36*, 1121–1124. [CrossRef]
20. Bragança, S.; Arezes, P.; Carvalho, M.; Ashdown, S.P.; Castellucci, I.; Leão, C. A comparison of manual anthropometric measurements with Kinect-based scanned measurements in terms of precision and reliability. *Work* **2018**, *59*, 325–339. [CrossRef]
21. *ISO 7250-1*; Basic Human Body Measurements for Technological Design—Part 1: Body Measurement Definitions and Landmarks. International Organization for Standardization: Geneva, Switzerland, 2017.
22. Castellucci, H.I.; Arezes, P.M.; Molenbroek, J.F.M. Applying different equations to evaluate the level of mismatch between students and school furniture. *Appl. Ergon.* **2014**, *45*, 1123–1132. [CrossRef]
23. Fidelis, O.P.; Ogunlade, B.; Adelakun, S.A. Incidence of School Furniture Mismatch and Health Implications in Primary School Children in Akure, South-West Nigeria. *J. Occup. Ther. Sch. Early Interv.* **2021**, *14*, 90–102. [CrossRef]
24. Alibegović, A.; Mačak Hadžiomerović, A.; Pašalić, A.; Domljan, D. School Furniture Ergonomics in Prevention of Pupils' Poor Sitting Posture. *Drv. Ind.* **2020**, *71*, 88–99. [CrossRef]
25. Prieto-Lage, I.; Ayán, C.; Alonso-Fernández, D.; Paramés-González, A.; Argibay-González, J.C.; Taboada-Solla, E.M.; Gutiérrez-Santiago, A. Degree of mismatch between anthropometric characteristics and school furniture in a sample of Spanish students aged 6–12 years old: A pilot study. *Arch. Argent. Pediatr.* **2021**, *119*, 386–393. [CrossRef] [PubMed]
26. Carrascosa, A. Aceleración secular de crecimiento en España. Estudios Españoles de Crecimiento 2010. Población autóctona y población inmigrante. *Endocrinol. Y Nutr.* **2014**, *61*, 229–233. [CrossRef]
27. Rexhepi, A.; Brestovci, B. Differences in growth and development velocity between boys and girls from Kosovo, aged 6–18 years. *Int. J. Med. Surg. Sci.* **2020**, *7*, 4–13. [CrossRef]
28. Castellucci, H.I.; Catalán, M.; Arezes, P.M.; Molenbroek, J.F.M. Evaluation of the match between anthropometric measures and school furniture dimensions in Chile. *Work* **2016**, *53*, 585–595. [CrossRef]

29. Aghayev, A.; Gurel, K.; Simsek, O.; Ipek, H.; Eser, E. Nutritional status and frequency of overweight and physical activity behaviors of medical school students and factors affecting these variables. *Med. Sci. | Int. Med. J.* **2021**, *10*, 64. [CrossRef]
30. Bravo, G.; Bragança, S.; Arezes, P.M.; Molenbroek, J.F.M.; Castellucci, H.I. A literature review of anthropometric studies of school students for ergonomics purposes: Are accuracy, precision and reliability being considered? *Work* **2018**, *60*, 3–17. [CrossRef]
31. Lee, Y.; Kim, Y.M.; Lee, J.H.; Yun, M.H. Anthropometric mismatch between furniture height and anthropometric measurement: A case study of Korean primary schools. *Int. J. Ind. Ergon.* **2018**, *68*, 260–269. [CrossRef]
32. Yanto; Lu, C.W.; Lu, J.M. Evaluation of the Indonesian National Standard for elementary school furniture based on children's anthropometry. *Appl. Ergon.* **2017**, *62*, 168–181. [CrossRef]
33. Cantin, N.; Delisle, I.; Baillargeon, M. Reducing Child-Furniture Incompatibility in Primary Schools. *J. Occup. Ther. Sch. Early Interv.* **2019**, *12*, 200–209. [CrossRef]
34. Castellucci, H.I.; Arezes, P.M.; Molenbroek, J.F.M. Analysis of the most relevant anthropometric dimensions for school furniture selection based on a study with students from one Chilean region. *Appl. Ergon.* **2014**, *46*, 201–211. [CrossRef]
35. Molenbroek, J.F.M.; Kroon-Ramaekers, Y.M.T.; Snijders, C.J. Revision of the design of a standard for the dimensions of school furniture. *Ergonomics* **2003**, *46*, 681–694. [CrossRef]

symmetry

MDPI

Article

The Magnitude of Temporal–Spatial Gait Asymmetry Is Related to the Proficiency of Dynamic Balance Control in Children with Hemiplegic Cerebral Palsy: An Analytical Inquiry

Ragab K. Elnaggar [1,2,†]

1 Department of Health and Rehabilitation Sciences, College of Applied Medical Sciences, Prince Sattam bin Abdulaziz University, Abdullah Ibn Amer St., Al-Kharj 11942, Saudi Arabia
2 Department of Physical Therapy for Pediatrics, Faculty of Physical Therapy, Cairo University, Giza 12613, Egypt
† rke_pt2001@yahoo.com; Abdullah Ibn Amer St., Al-Kharj 11942, Saudi Arabia.

Abstract: Children with hemiplegic cerebral palsy (hemi-CP) frequently experience deficits in dynamic balance, a crucial factor influencing gait function. This imbalance can manifest as temporal–spatial gait asymmetry, where movement patterns differ between the affected and less affected sides. This study investigated how temporal–spatial gait asymmetries and dynamic balance are associated in children with hemi-CP. Eighty-five children with hemi-CP (age: 13.27 ± 1.72 years) were included. The temporal ($AI_{Temporal}$) and spatial ($AI_{Spatial}$) gait asymmetry indices were, respectively, computed with reference to the swing time and step length of affected and less affected sides, which were collected through a 3D gait analysis. Measures of dynamic balance included the directional dynamic limit-of-stability ($D\text{-}LOS_{directional}$) assessed across multiple directions (forward, rearward, affected, and less affected) and the overall dynamic limit-of-stability ($D\text{-}LOS_{overall}$) during static stance, in addition to the heel-to-heel base of support ($BOS_{H\text{-}to\text{-}H}$) during walking, the dynamic gait index (DynGI), and the Timed Up and Down Stair (TUDS) test.The $D\text{-}LOS_{overall}$ correlated negatively with the temporal ($r = -0.437$, $p < 0.001$) and spatial ($r = -0.279$, $p = 0.009$) asymmetries. The $D\text{-}LOS_{directional}$ (forward, rearward, affected, and less affected) correlated negatively with temporal asymmetry (r ranged from -0.219 to -0.411, all $p < 0.05$), but only the $D\text{-}LOS_{directional}$ rearward ($r = -0.325$, $p = 0.002$) and less affected ($r = -0.216$, $p = 0.046$) correlated with spatial asymmetry. The $BOS_{H\text{-}to\text{-}H}$ correlated positively with both temporal ($r = 0.694$, $p < 0.001$) and spatial ($r = 0.503$, $p < 0.001$) asymmetries. The variation in $D\text{-}LOS_{overall}$ and $BOS_{H\text{-}to\text{-}H}$ accounted for 19.1% and 48.2%, respectively, of the variations in the temporal asymmetry and 7.8% and 25.3% of the variations in the spatial asymmetry. The findings of this study suggest that dynamic balance control is related to the magnitude of temporal–spatial gait asymmetries in children with hemi-CP. This evidence lays the groundwork for further research into the mechanism linking gait asymmetry and dynamic balance, potentially leading to a deeper understanding of these impairments, while also highlighting the need for longitudinal studies with the inclusion of a broader population to enhance the generalizability of the findings.

Keywords: cerebral palsy; gait symmetry index; gait analysis; dynamic balance control; base of support

1. Introduction

Cerebral palsy (CP) is a group of permanent disorders related to the development of movement and posture, which results in activity limitations. These disorders arise from non-progressive disturbances in the developing brain during the fetal or early infancy stages. These motor disorders frequently coexist with sensory, perceptual, cognitive, and behavioral disturbances in sensation, perception, cognition, as well as behavior, in addition to epilepsy and secondary musculoskeletal issues [1]. The estimated prevalence rate of CP in high-income countries is 2.1 per 1000 live newborns [2], while the exact rate in middle and low-income countries remains uncertain, but appears to be higher, with greater physical disability [3]. Hemiplegic CP (hemi-CP) is a frequent type of CP in which motor deficits are predominant on one side [4]. This type accounts for 33–39% of all CP occurrences [5].

Children with hemi-CP frequently present with distinctive gait deviations that hinder their mobility and daily functioning. Winter's classification categorizes these deviations based on sagittal plane kinematics, each reflecting unique functional challenges. Type I is marked as a "drop foot" during the swing phase, while Type II features a true equinus due to spasticity of the gastrocnemius muscles. Type III presents a stiff knee gait resulting from co-contraction of the hamstrings and quadriceps, and Type IV shows more pronounced proximal involvement, resembling patterns seen in spastic diplegia [6]. A thorough understanding of the interplay between these gait deviations and other clinical impairments—such as muscle weakness, spasticity, restricted range of motion, loss of selective motor control, and impaired balance control—is crucial. This knowledge can guide clinicians in developing targeted management strategies that effectively address the multifaceted challenges these children encounter.

Efficient and safe locomotion is the overriding objective of gait training in children with hemi-CP [7,8]. Enhancing gait function ensures higher levels and diversity of activity/participation in their daily life [9,10]. Children with hemi-CP, even after rehabilitation, continue to have spasticity, weakness, and abnormal patterns of muscle activation, all of which have been shown to restrict children's mobility and impede them from fully participating in daily activities [11]. In addition, children with hemi-CP experience a pronounced asymmetry of body weight distribution between lower limbs during quiet standing. They become more reliant on the less affected limb compensating for the weakness of the affected side [12]. The propensity to keep the body weight shifted on the less affected limb is also seen during ambulation [13–16]. As a consequence, children may evolve step-length and limb-phasing asymmetries [17]. The ensuing asymmetries are specifically associated with slow, inefficient walking and poor balance control [15,18].

Earlier research has sought to pinpoint key factors influencing gait asymmetry in children with hemi-CP [19,20]. Several impairments, including spasticity, muscle weakness/stiffness, and impaired proprioception, have been marked among the major determinants of gait asymmetries [19]. The latest research suggested that in addition to kinematic deviations of the affected lower limb [20], reduced balance capability during quiet standing is associated with asymmetrical gait patterns [16,18]. Despite the fact that quiet standing balance and gait are two distinct levels of motor functions, they have integrated control mechanisms (i.e., they share some organizational principles, such as the control of the center of mass and the frame of reference for their respective kinematic coordination). Additionally, standing balance and gait have a number of associations and interdependencies at different CNS levels—that is, the afferent feedback serves as a mediator in the relationship between balance and gait regulation and many neural pathways operate to shape the pattern of CNS responses [21].

The foregoing analyses were limited to studying the relationship between static balance and gait asymmetries, where instead of using data from a computerized instrument, the balance was determined by the degree of weight distribution towards the less affected side or a clinical balance scale [16,22]. This, unlike dynamic balance data that are obtained via computerized equipment, does not involve the dynamic limits of postural control—a direction-specific shifting of the body weight (i.e., forward/rearward or the affected/less affected side direction) during standing and/or dynamic balance control in functional contexts such as walking and stair climbing. The earlier findings, therefore, may fall short of fully explaining the link between gait asymmetries and dynamic balance.

A nuanced understanding of the relationship between multifarious balance metrics and gait function is imperative to establish a credible rationale for targeted rehabilitation strategies that enhance functional mobility [23], ultimately helping physical rehabilitation practitioners make informed treatment decisions. Accordingly, this study sought to determine whether an association exists between temporo-spatial gait asymmetries and dynamic balance control both during standing and during walking—in a convenience sample of children with hemi-CP. The initial hypothesis was that gait asymmetries would be associated with reduced dynamic balance control.

2. Materials and Methods

2.1. Study Protocol and Ethics

This was an analytical inquiry undertaken between December 2019 and August 2021 at Biomechanics Laboratories (Gait/Motion Analysis Lab and Balance Assessment Lab) of the Department of Physical Therapy at Prince Sattam Bin Abdulaziz University, Al-Kharj, KSA. Participants and their parents/legal representatives were informed about the study's objective and procedures before joining the study and signing a written consent form. The study protocol was approved by the Physical Therapy Research Ethics Committee at the university (RHPT/0019/0040). Procedures conformed with the ethical guidelines of the Declaration of Helsinki, which was released in 1975.

2.2. Participants

Participants were recruited via the Neurology/Physical Therapy departments of four local hospitals in Riyadh/Al-Kharj, KSA. Attending clinicians who normally see children with CP in the outpatient clinics of these hospitals were given a thorough explanation of the study so that they could make appropriate referrals. Participants were further screened for eligibility by the principal investigator, who had more than 20 years of experience in pediatric physical therapy. Inclusion criteria were: (1) a pediatric neurologist-verified hemi-CP diagnosis [24]; (2) age between 8–15 years (since walking behavior in this age group tends to be biomechanically more consistent with the adult patterns) [19,25]; (3) motor function level I or II per the gross motor function classification system (GMFCS) [26], representing children with mild-to-moderate motor impairments who are capable of independent ambulation, which allows for investigation of gait asymmetry and dynamic balance in more active and mobile group; (4) mild spasticity, that is, a spasticity grade of 1 or 1+ on the modified Ashworth scale (MAS) [27]—assessed through hip adductors, knee extensors and flexors, and ankle plantar flexors—to minimize variability and potential confounding factors associated with the more severe spasticity; and (5) adequate mental capacity—this was verified if the children's records contained pertinent evidence proving normal intellectual function, if they were capable of perceiving, comprehending, and following instructions effectively during screening, and if they were enrolled in regular traditional school classes where they could focus, interact, and work collaboratively with others. Exclusion criteria were neurolytic blocking agents in the past six months, corrective musculoskeletal surgery through the past year, severe contractures, leg-length discrepancy, and visual/auditory deficits.

Power Analysis and Sample Size Determination

The sample size was determined depending on a 95% confidence interval (95%CI: 0.275–0.646) for a Pearson correlation coefficient representing the relationship between dynamic balance and temporal asymmetry index. These were obtained by analyzing data from the first 12 observations. A sample size of 68 children was required to produce a two-sided 95%CI with a width equal to 0.371 when the estimate of Pearson's product–moment correlation was (r = 0.482). However, this study collected data from 85 children, expecting that ~20% of participants might be lost at random. The power analysis was performed using the PASS software, v16.0.12 (NCSS, Kaysville, UT, USA).

2.3. Measurements

Each child participated in two measurement sessions held on two consecutive days. The first session involved 3D gait analysis. The second session included a laboratory-based assessment of dynamic balance, which utilized the Biodex Balance System to determine both the directional and overall dynamic limits of postural stability. This was followed by an evaluation of clinical metrics related to dynamic balance, specifically the timed up and down stairs test and dynamic gait index. A 15- to 20-min rest period was incorporated between these measurements.

2.3.1. Computation of Gait Asymmetry

The 3D VICON MX motion-capture system (Oxford Metrics Ltd., Oxford, UK) was used to collect gait data at a 1200 Hz sampling frequency. Twelve near-infrared cameras (4-megapixel resolution, maximum speed of 370 frames/second) were employed to track 17 light-reflective markers during walking in a 3D coordinate system within a 1.5 mm spatial displacement error. The system was calibrated, and the projection area was tuned as part of the measurement preparation. Markers were affixed to the skin overlaying particular bone landmarks, as previously outlined [20,28]. Participants were instructed to walk barefoot five times along a 10-m walking route at a comfortable pace. Data were processed using the VICON Polygon software (v4.3.2). To minimize the acceleration and deceleration effects, three usable gait cycles were picked up from the middle of the walking route. The average temporo-spatial gait parameters were computed. The spatial asymmetry index ($AI_{Spatial}$) was computed for the step length of the affected ($StepL_{AFF}$) and less affected ($StepL_{Less\text{-}AFF}$) sides as follows: $AI_{Spatial}$ = abs $[(StepL_{AFF} - StepL_{Less\text{-}AFF})/[(StepL_{AFF} + StepL_{Less\text{-}AFF})]$. The temporal asymmetry index ($AI_{Temporal}$) was computed for the swing times of the affected ($SwingT_{AFF}$) and less affected ($SwingT_{Less\text{-}AFF}$) sides as follows: $AI_{Temporal}$ = abs $[(SwingT_{AFF} - SwingT_{Less\text{-}AFF})/[(SwingT_{AFF} + SwingT_{Less\text{-}AFF})]$ [22]. Higher values mean more asymmetrical patterns, and a value of "zero" denotes symmetrical patterns [19,20].

2.3.2. Quantification of Dynamic Balance

The dynamic limit of stability (D-LOS) during static stance was quantified using the Biodex balance system (Biodex Medical Systems, Shirley, NY, USA). The D-LOS designates the maximum excursion of the center of gravity (COG) that children were deliberately able to cover in different directions through the base of support (BOS) without stepping or losing balance. In the D-LOS test, the transfer of the COG to intercept eight consecutive targets arising randomly at intervals of 45° around the center of pressure was assessed in terms of timing and accuracy. These targets, according to the manufacturer, emerge at 50% of the maximum possible COG excursion, which depends on each child's height.

Participants stood on the balance plate in a natural stance, barefoot, with their feet hip-width apart and their arms at their sides. Then, they were directed to lean their bodies to the furthest extent to move the COG (represented by a cursor on a display panel) toward a target, and then to return to the center before the next target was displayed. Children were taught to perform the test as accurately and quickly as was feasible while maintaining a straight body and employing their ankles as key axes of movement. Throughout the test, children were observed by the assessor to ensure that the motions were dominant at their ankles and there were no wide segmental motions at the trunk, hip, or knee. If the test revealed noticeably increased COG trajectories (the biomechanical expression of large segmental motion), it was repeated to make sure of the correct execution of the test without segmental motions. Once the eight targets had been accomplished, the test was terminated. Three trials were allowed, and the average score was recorded. For the purpose of this study, the $D\text{-}LOS_{directional}$ (the average directional control score for the forward, rearward, affected, and less affected directions only) and the $D\text{-}LOS_{overall}$ were used for statistical computations. Higher D-LOS scores indicate better dynamic balance control. Below are the algorithms for computing the $D\text{-}LOS_{directional}$ and $D\text{-}LOS_{overall}$ scores [14,29,30].

$$\text{D-LOS}_{\text{directional}}(\%) = \frac{\text{Rectilinear distance to target}}{\text{Entire distance covered}} \times 100$$

$$\text{D-LOS}_{\text{overall}}\ (\%) = \sum_{i=1}^{i=4} (\text{D} - \text{LOS}_{\text{directional}}\ \text{score}) \div 4\ (\text{Mean of four targets})$$

In addition, the heel to heel BOS ($BOS_{H\text{-to-}H}$; the perpendicular distance between the heel center of one footfall at the affected side and the progression line formed by two footfalls on the less affected side) [30,31], as one of the variables measured through gait analysis, was employed to assess the ability to maintain dynamic balance during walking. A larger $BOS_{H\text{-to-}H}$ value indicates a lower dynamic balance competence.

2.3.3. Clinical Metrics of Dynamic Balance

The dynamic gait index (DynGI) was used to evaluate the dynamic balance control and risk of falling while walking. It is a valid, performance-based, and simple-to-use instrument for assessing the children's potential to modify their balance, not only during habitual steady-state walking but also for walking during more challenging tasks (i.e., in the presence of external demands), in children from 8 to 15 years old [32]. The DynGI is composed of eight items that are assessed on a four-point ordinal scale (0–3, with 0 representing severe impairment and 3 indicating normal performance). The DynGI items include the following: leveled-surface walking, walking pace change, horizontal and vertical head turns, stepping over and around obstacles, walking and pivot turns, and stepping up/down stairs. The maximum DynGI score is 24, which indicates that the child is able to walk safely [32].

Another clinical metric for assessing dynamic balance capabilities was the Timed Up and Down Stairs (TUDS) test. The TUDS test has been demonstrated to be valid and reliable in children with typical development and children with CP whose motor function is classified as level I or II on the GMFCS [33]. The test was conducted on a flight of stairs with 14 steps (each measuring 20 cm in height). For testing, children stood up 30 cm away from the bottom step. They were then instructed to climb up the stairs as quickly as they felt safe, turn around a mark on the top of the stairs, and go all the way downstairs until both feet hit the starting point. Children were free to use any stair-climbing strategy they wanted (examples of these are, skipping steps, running up the stairs, using a step-to or foot-over-foot sequence, or others), but were advised to face the moving direction (that is; to face up and down as they traverse the steps, rather than to the side). They were given a signal "ready, 1, 2, 3, and go". The tests were conducted while wearing shoes, but not lower limb orthotics. The time (second) between the "go" signal and the landing was measured through a stopwatch. Shorter times suggest a higher performance level.

2.4. Data Analysis

The NCSS Statistical Program for Windows, version 11.0.13 (NCSS Statistical Software©, Kaysville, UT, USA) was used for all statistical analyses, and GraphPad PRISM 9 (GraphPad Software Inc.; San Diego, CA, USA) was used for graph generation. The evidence against the null hypothesis was indicated by a p-value of 0.05 or lower. Descriptive statistics (mean ± StDev and min/max observations) were computed to provide a comprehensive overview of the key characteristics. The Pearson correlation coefficient was used to characterize the degree/direction of the relationship between dynamic balance and temporo-spatial gait asymmetries. Linear regression analysis was performed to determine how dynamic balance affected gait asymmetries.

3. Results

Eighty-five children completed the required measurements, and their data were included in the analysis. Demographic and clinical characteristics of the participating children are outlined in Table 1, whereas the descriptive statistics related to gait and dynamic balance measurements are summarized in Table 2.

Table 1. Characteristics (demographic, anthropometric, and clinical) of the participating children.

Variables	Values	Min–Max
Age, years	13.27 ± 1.72	8–15
Gender (boys/girls), *n* (%)	54 (63.5)/31 (36.5)	NA
Height, m	1.39 ± 0.10	1.19–1.61
Weight, Kg	41.87 ± 6.53	30–61
BMI. Kg/m^2	21.41 ± 1.34	18.97–24.13
Side affected (RT/LT), *n* (%)	39 (45.9)/46 (54.1)	NA
Spasticity per MAS (1/1+), *n* (%)	48 (56.5)/37 (43.5)	NA
GMFCS level (I/II), *n* (%)	57 (67.1)/28 (32.9)	NA
Insult/MalDev site (C/SC/C+SC), *n* (%)	19 (22.3)/56 (65.9)/10 (11.8)	NA
Insult type (WMI/GMI/B-MalDev), *n* (%)	53 (62.4)/23 (27.1)/9 (10.6)	NA
AFO use (yes/no), *n* (%)	29 (34.1)/56 (65.9)	NA

Age, height, weight, and BMI are shown as mean ± standard deviation. Other variables are listed as frequency (percentage). Abbreviations: RT/LT: right/left side, MAS: Modified Ashworth Scale, GMFCS: gross motor function classification system, MalDev: maldevelopment, C: cortical, SC: subcortical, WMI: white matter insult, GMI: grey matter insult, B-MalDev: brain maldevelopment, AFO: ankle–foot orthosis, NA: not applicable.

Table 2. The temporo-spatial gait parameters, asymmetry indices, and dynamic balance variables in the participating children.

Variable		Mean ± StDev	95% LCL, UCL for Mean	Min–Max	*p*-Value
Temporo-Spatial Gait Parameters					
Walking speed, cm/s		92.13 ± 15.25	88.84–95.42	48.11–120.42	NA
StepL, cm	Affected	0.62 ± 0.05	0.61–0.63	0.48–0.71	<0.001 **
	Less affected	0.49 ± 0.04	0.48–0.50	0.39–0.57	
SwingT, s	Affected	0.49 ± 0.06	0.47–0.50	0.34–0.64	<0.001 **
	Less affected	0.38 ± 0.05	0.37–0.39	0.29–0.51	
Asym. indices	$AI_{Spatial}$	0.11 ± 0.04	0.10–0.12	0.03–0.19	NA
	$AI_{Temporal}$	0.12 ± 0.05	0.11–0.13	0.01–0.23	NA
Dynamic balance measures					
D-$LOS_{directional}$	Forward	43.88 ± 5.46	42.70–45.10	27–58	NA
	Rearward	42.51 ± 6.21	41.17–43.85	28–54	NA
	Affected	46.76 ± 8.04	45.03–48.50	27–65	NA
	Less affected	54.68 ± 6.32	53.32–56.05	41–68	NA
D-$LOS_{overall}$		46.96 ± 4.25	46.04–47.87	34–59	NA
$BOS_{H\text{-}to\text{-}H}$, cm		17.31 ± 4.69	16.24–18.38	6.51–26.50	NA
DynGI		18.67 ± 2.12	18.21–19.13	14–22	NA
TUDS, s		17.12 ± 4.55	16.14–18.10	8.5–27.4	NA

Abbreviations: StepL: step length, SwingT: swing time, Asym: asymmetry, AISpatial: spatial asymmetry index, AITemporal: temporal asymmetry index, D-LOS: dynamic limit of stability, BOSH-to-H: heel-to-heel base of support, DynGI: dynamic gait index, TUDS: timed up and down stair test, NA: not applicable. StDev: standard deviation, LCL: lower confidence limit, UCL: upper confidence limit, *p*-value indicates the difference between affected and non-affected sides, ** significant at $p < 0.01$.

The analysis uncovered a significant negative correlation between the $AI_{Spatial}$ and some measures of the D-$LOS_{directional}$, specifically the rearward [$r = -0.325$; $p = 0.002$] and less affected direction [$r = -0.216$; $p = 0.046$] and the D-$LOS_{overall}$ [$r = -0.279$; $p = 0.009$]. Furthermore, there was a significant positive correlation between $AI_{Spatial}$ and $BOS_{H\text{-}to\text{-}H}$ [$r = 0.503$; $p < 0.001$] and TUDS duration [$r = 0.265$; $p = 0.014$] and a negative correlation between $AI_{Spatial}$ and DynGI [$r = -0.280$; $p = 0.009$] (Table 3).

Table 3. Associations between dynamic balance metrics and temporo-spatial gait asymmetry indices.

	$AI_{Spatial}$		$AI_{Temporal}$	
	r (95%CI)	*p*-Value	r (95%CI)	*p*-Value
$D\text{-}LOS_{directional}$				
Forward	−0.204 (0.009, −0.399)	0.059	−0.219 (−0.007, −0.411)	0.043 *
Rearward	−0.325 (−0.119, −0.501)	0.002 **	−0.411 (−0.216, −0.572)	<0.001 **
Affected side	−0.028 (0.185, −0.239)	0.797	−0.272 (−0.062, −0.456)	0.012 *
Less affected side	−0.216 (−0.119, −0.501)	0.046 *	−0.243 (−0.031, −0.432)	0.025 *
$D\text{-}LOS_{overall}$	−0.279 (−0.003, −0.409)	0.009 **	−0.437 (−0.245, −0.592)	<0.001 **
$BOS_{H\text{-}to\text{-}H}$	0.503 (0.323, 0.645)	<0.001 **	0.694 (0.563, 0.791)	<0.001 **
DynGI	−0.280 (−0.071, −0.463)	0.009 **	−0.297 (−0.089, −0.478)	0.006 **
TUDS	0.265 (0.055, 0.451)	0.014 *	0.214 (0.001, 0.406)	0.049 *

Abbreviations: AISpatial: spatial asymmetry index, AITemporal: temporal asymmetry index, D-LOS: dynamic limit of stability, BOSH-to-H: heel-to-heel base of support, DynGI: dynamic gait index, TUDS: timed up and down stair test, r: Pearson correlation coefficient, CI: confidence intervals. * Significant at $p < 0.05$, ** significant at $p < 0.01$.

A significant negative correlation was observed between $AI_{Temporal}$ and all measures of the $D\text{-}LOS_{directional}$ [forward: $r = -0.219$; $p = 0.043$, rearward: $r = -0.411$; $p < 0.001$, affected: $r = -0.272$; $p = 0.012$, and less affected direction: $r = -0.243$; $p = 0.025$], as well as the $D\text{-}LOS_{overall}$ [$r = -0.437$; $p < 0.001$]. There was also a significant positive correlation between the $AI_{Temporal}$ and the $BOS_{H\text{-}to\text{-}H}$ [$r = 0.694$; $p < 0.001$] and TUDS duration [$r = 0.214$; $p = 0.049$], as well as a negative correlation between $AI_{Temporal}$ and DynGI [$r = -0.297$; $p = 0.006$] (Table 3).

The estimated change in $AI_{Spatial}$ per unit change in $D\text{-}LOS_{overall}$ was −0.0024. The $D\text{-}LOS_{overall}$ accounted for ~7.8% of variations in the $AI_{Spatial}$ ($R^2 = 0.0776$). The value of $AI_{Spatial}$ when the $D\text{-}LOS_{overall}$ was zero was 0.2250. The straight-line equation relating $AI_{Spatial}$ and $D\text{-}LOS_{overall}$ was estimated as: $AI_{Spatial} = (0.2250) + (-0.0024)\ D\text{-}LOS_{overall}$ (Figure 1a). The analysis also revealed that the estimated change in $AI_{Spatial}$ per unit change in $BOS_{H\text{-}to\text{-}H}$ was 0.0038. The proportion of variation in $AI_{Spatial}$ that can be explained by the variation in $BOS_{H\text{-}to\text{-}H}$ was ~25.3% ($R^2 = 0.2533$). The value of $AI_{Spatial}$ when the $BOS_{H\text{-}to\text{-}H}$ was zero was 0.0448. The straight-line equation relating $AI_{Spatial}$ and $D\text{-}LOS_{overall}$ was estimated as: $AI_{Spatial} = (0.0448) + (0.0038)\ BOS_{H\text{-}to\text{-}H}$ (Figure 1b).

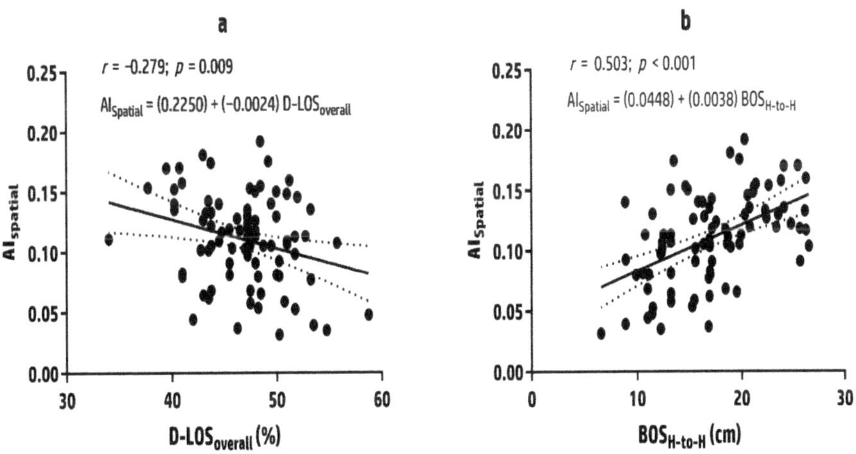

Figure 1. The regression lines relating spatial gait asymmetry index ($AI_{Spatial}$) to the overall dynamic limit of stability ($D\text{-}LOS_{overall}$) and heel-to-heel base of support ($BOS_{H\text{-}to\text{-}H}$). The continuous line depicts the mean slope of the group-derived regression line while the dotted lines represent the standard error of the mean regression line.

The estimated change in $AI_{Temporal}$ per unit change in D-$LOS_{overall}$ was –0.0050. The D-$LOS_{overall}$ accounted for ~19.1% of variations in the $AI_{Temporal}$ (R^2 = 0.1909). The y-intercept, the value of $AI_{Temporal}$ when the D-$LOS_{overall}$ is zero, was 0.3582. The regression-line equation relating $AI_{Temporal}$ and D-$LOS_{overall}$ was estimated as: $AI_{Temporal}$ = (0.3582) + (–0.0050) D-$LOS_{overall}$ (Figure 2a). The estimated change in $AI_{Temporal}$ per unit change in $BOS_{H\text{-to-}H}$ was 0.0069. The proportion of variation in $AI_{Temporal}$ that could be explained by the variation in $BOS_{H\text{-to-}H}$ was ~48.2% (R^2 = 0.4821). The value of $AI_{Temporal}$ when the $BOS_{H\text{-to-}H}$ was zero was 0.0036. The regression-line equation relating $AI_{Temporal}$ and $BOS_{H\text{-to-}H}$ was estimated as: $AI_{Temporal}$ = (0.0036) + 0.0069) $BOS_{H\text{-to-}H}$ (Figure 2b).

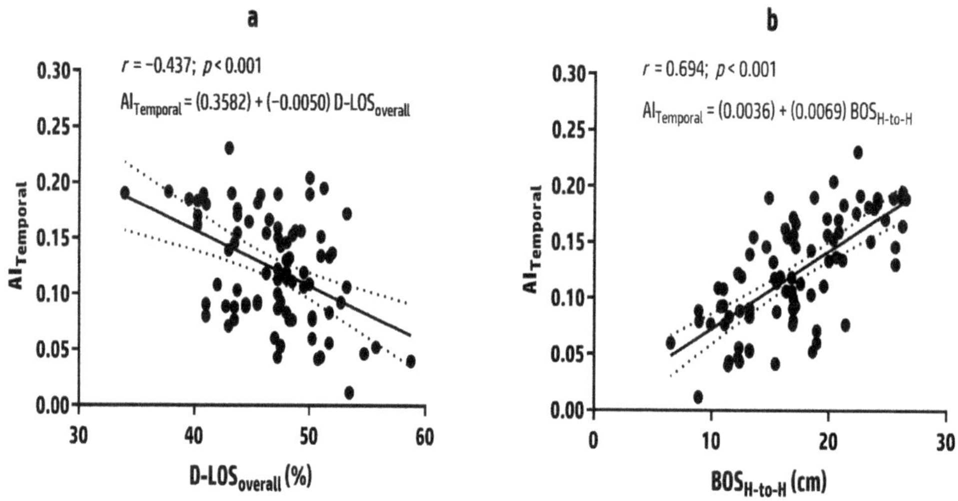

Figure 2. The trend lines relating temporal gait asymmetry index ($AI_{Temporal}$) to the overall dynamic limit of stability (D-$LOS_{overall}$) and heel-to-heel base of support ($BOS_{H\text{-to-}H}$). The continuous line depicts the mean slope of the group-derived regression line while the dotted lines represent the standard error of the mean regression line.

4. Discussion

The fundamental objective of the current work was to determine if an association existed between the proficiency of dynamic balance control (on standing and during walking) and the asymmetry patterns of gait (spatial and temporal) in children with hemi-CP. The most evident finding to emerge from this analytical study is that the dynamic balance control in a static stance or during walking is significantly related to the magnitude of temporo-spatial asymmetries of gait. More specifically, higher D-LOS scores (i.e., shifting body weight in different directions more efficiently) during static stance, narrower $BOS_{H\text{-to-}H}$ distance during walking, lower DynGI score, and shorter TUDS duration are associated with the more symmetrical patterns. These findings substantiate the initial hypothesis that gait asymmetries are intricately related to diminished dynamic balance control in children with hemi-CP.

It is generally recognized that children with hemi-CP have variable degrees of postural control and balance impairments, which are key elements of locomotor issues [34]. However, no research exists exploring the relationship between dynamic balance and temporospatial gait asymmetries. Children with hemi-CP tend preferentially to support more body weight on the less affected lower limb while standing and during walking [12]. This differential loading resulted in decreased balance control of the affected limb and was intrinsically linked to the deteriorated temporo-spatial asymmetries of gait in such a patient population [16,17].

The current findings revealed that temporal gait asymmetry decreases among children who have a greater capacity for adjusting their COG within a fixed BOS in both forward/rearward and affected/less affected side directions without losing their balance. Children with hemi-CP mostly have trouble shifting their weight toward the affected lower limb, resulting in more instability during the stance phase, and leading eventually to a shorter swing phase on the less affected side [12,16,18]. Nevertheless, the capacity to transfer body weight in antero-posterior and lateral directions involves the affected side's direction and permits further stance stability during walking; therefore, the less affected lower limb's swing phase is lengthened and temporal asymmetry is diminished. These findings are further corroborated by the evidence from a preceding analysis which pointed out that an augmented weight-bearing capability of the affected limb promotes the enhancement of the swing time symmetry [35].

It has also been observed, surprisingly, that spatial gait asymmetry is lower in children who are able to shift their COG than it is in children who are unable to shift their COG toward specific directions (rearward or less affected sides' direction) rather than others. Seemingly, different factors like ankle plantar flexors' spasticity, plantar and dorsiflexors' swing trajectory contribution, and kinematic deviations of the lower limbs have greater associations with the spatial asymmetry of gait (although not directly measured in this study) and play even more important roles in determining the spatial asymmetry patterns than the ability to maintain dynamic balance (just as body weight shifting) [19,20]. The plantar flexor spasticity brings on further instability during the stance phase and inefficient weight support on the affected limb. Therefore, children with hemi-CP are likely to reduce the $StepL_{Less\text{-}AFF}$ in comparison with the $StepL_{AFF}$ [19]. Insufficient strength of the ankle joint muscles could be linked to the shortened $StepL_{AFF}$. The ankle dorsiflexors prevent foot dragging along the ground and inhibit premature foot contact [36], therefore extending the StepL directly, whereas the plantar flexors act during the pre-swing phase to lengthen the StepL by creating the ground reaction force [36,37]. Thus, factors that directly influence the StepL, like the spasticity of the plantar flexors and the strength of the dorsi and plantar flexors, have stronger associations with the spatial asymmetry of gait than dynamic balance control during static standing [19].

The current study determined that the temporo-spatial gait asymmetries decreased in children who exhibited narrower $BOS_{H\text{-}to\text{-}H}$ distributions, greater DynGI scores, and shorter TUDS durations. That is, children with more proficient dynamic balance functions were capable of walking within a narrow BOS and navigating stairs more quickly, implying that they were able to effectively maintain their balance during either walking or performance of other locomotor activities by shifting their COG in all forward/rearward and affected/less affected directions. These findings are in accordance with prior studies on adults with post-stroke hemiplegia, which found that patients who experience more temporo-spatial gait asymmetries walk with a wider step, which is linked to decreased balance competencies [30,38].

4.1. Study Merits and Literature Contribution

This study constitutes a groundbreaking investigation into the complex interplay between temporo-spatial gait asymmetries and dynamic balance control specifically within the context of the pediatric population with hemi-CP. To the author's knowledge, it represents the first comprehensive exploration of this critical relationship, thereby offering novel insights and contributing significantly to the existing body of literature. While previous studies have examined these variables independently [13,34,39–41], the findings demonstrated herein underscore the interrelatedness of the gait mechanics and balance, highlighting that impaired dynamic balance can significantly impact the pattern of gait symmetry. Additionally, this study utilized a relatively large sample and a high power (95%), crucial elements that allow for the detection of smaller but meaningful relationships and significantly enhance the reliability and validity of the findings. Moreover, the current findings advocate for the implementation of integrated assessment approaches in clinical practice, where both gait asymmetries and dynamic balance deficits are evaluated concurrently. This holistic perspective is essential for informing targeted therapeutic interventions that address both aspects simultaneously.

4.2. Study Limitations

Despite the significance of the findings demonstrated herein, certain limitations may apply to their generalizability. The paucity of comparable results is one source of uncertainty. Therefore, further research into the relationship between gait symmetry and dynamic balance control in children with hemi-CP would contribute to developing a higher level of precision in this issue. The study included children whose brain insult/mal-development site was heterogeneous (i.e., cortical, subcortical, or a combination of both), and temporo-spatial parameters of gait might differ from one another. Therefore, studies with larger samples stratified by the brain insult/mal-development site are warranted to affirm or disprove the current findings. The study analyzed data from children within the 8–15 years age bracket. So, the question of whether children with hemi-CP in younger or older age groups would have similar outcomes remains unanswered, and this could be a starting point for further analyses in the future. A note of caution is also due here since the sample was restricted to children with mild spasticity (i.e., level 1 or 1+ on the MAS), which may make the findings less generalizable to all children with hemi-CP, especially as some of those children could have more spasticity in at least some of the joints (hamstrings, rectus, or gastrocnemius) in the lower extremities. Therefore, additional work is required to confirm the viability of the current findings for children with higher spasticity. The current findings should also be interpreted with caution given the fact that some children (about 1/3) were assessed for gait and dynamic balance barefoot, although they occasionally wore an ankle–foot orthosis. In future investigations, it might be possible to collect/analyze data with and without the use of orthosis, thereby drawing a more definitive conclusion. Since no reference values exist for the dynamic balance measures employed in the present study, it was not possible to indicate what was expected from children aged 8–15 and functioning at GMFCS level I or II; this could, therefore, be a line of inquiry for upcoming studies. Finally, the proprioceptive function was not considered when children were selected to join the study. Gait asymmetries have been observed to worsen with increasing degrees of position-sense error [13]. Thus, forthcoming research should consider this factor, thereby providing more definitive evidence.

4.3. Clinical and Research Implications

The study's findings on the relationship between gait asymmetry and dynamic balance can guide clinicians and physical rehabilitation practitioners in developing targeted therapeutic interventions. An in-depth understanding of specific gait patterns that correlate with balance deficits would help therapists create individualized rehabilitation programs that address both aspects simultaneously, enhancing overall mobility outcomes. Further, the insights from this study can lead to the establishment of standardized assessment protocols that evaluate gait asymmetry and dynamic balance, allowing for more accurate monitoring of patient progress and timely adjustments to the treatment plans. Furthermore, identifying specific gait asymmetries that predispose children to balance issues also enables the implementation of preventive strategies, ensuring early intervention before more significant mobility challenges arise.

This study also provides a foundation for future research to explore the underlying mechanisms linking gait asymmetry and dynamic balance in greater depth. Elucidation of these mechanisms could lead to the design of targeted interventions that address identified deficits. Prospective cohort studies could facilitate the monitoring of gait asymmetry and dynamic balance, contributing to a better understanding of the long-term trajectory of these impairments in children with hemiplegic cerebral palsy. Longitudinal studies are also warranted on how changes in gait asymmetry and dynamic balance evolve with various interventions, offering insights into their effectiveness and long-term impact. Additionally, expanding the research to include broader populations with varying severity levels of hemiplegic cerebral palsy would enhance the generalizability of the findings, informing more inclusive guidelines.

5. Conclusions

This study highlights a relationship between the magnitude of temporal–spatial gait asymmetry and dynamic balance control in children with hemi-CP. Generally, children exhibiting greater dynamic balance control tend to demonstrate more symmetrical patterns. Specifically, the D-LOS$_{directional}$ data suggest that increasing difficulties in shifting weight toward the affected limb are associated with increased gait asymmetries. However, the observed correlations were weak to moderate, suggesting that while these relationships were notable, they warrant further investigations to fully understand the complexities involved, particularly how dynamic balance control impacts gait symmetry in this population. Future investigations are, therefore, recommended to delve into the causal pathway linking gait asymmetry and dynamic balance, as this could deepen our understanding of their interplay in children with hemi-CP.

Funding: The authors extend their appreciation to Prince Sattam bin Abdulaziz University for funding this research work through the project number (PSAU/2024/01/78915).

Institutional Review Board Statement: The study protocol was approved by the Physical Therapy Research Ethics Committee at the university (RHPT/0019/0040). All procedures conformed with the ethical guidelines of the Declaration of Helsinki.

Informed Consent Statement: Informed consent was obtained from all participants who took part in the present study.

Data Availability Statement: The data that support the findings of this study are available on reasonable request from the corresponding author.

Acknowledgments: The author of this work is immensely grateful to the children and their families for their enthusiastic participation, as their cooperation and commitment were instrumental in the successful execution of this study.

Conflicts of Interest: The author declares no conflicts of interest.

References

1. Rosenbaum, P.; Paneth, N.; Leviton, A.; Goldstein, M.; Bax, M.; Damiano, D.; Dan, B.; Jacobsson, B. A report: The definition and classification of cerebral palsy April 2006. *Dev. Med. Child Neurol. Suppl.* **2007**, *109* (Suppl. S109), 8–14. [PubMed]
2. Oskoui, M.; Coutinho, F.; Dykeman, J.; Jetté, N.; Pringsheim, T. An update on the prevalence of cerebral palsy: A systematic review and meta-analysis. *Dev. Med. Child Neurol.* **2013**, *55*, 509–519. [CrossRef] [PubMed]
3. Blair, E.; Watson, L. Epidemiology of cerebral palsy. *Semin. Fetal Neonatal. Med.* **2006**, *11*, 117–125. [CrossRef]
4. Zerbino, L.; Giannoni, P. *The Child with Unilateral Spastic Cerebral Palsy. Cerebral Palsy: A Practical Guide for Rehabilitation Professionals*; Springer: Berlin/Heidelberg, Germany, 2022; pp. 127–171.
5. Jonsson, U.; Eek, M.N.; Sunnerhagen, K.S.; Himmelmann, K. Cerebral palsy prevalence, subtypes, and associated impairments: A population-based comparison study of adults and children. *Dev. Med. Child Neurol.* **2019**, *61*, 1162–1167. [CrossRef] [PubMed]
6. Rodda, J.; Graham, H.K. Classification of gait patterns in spastic hemiplegia and spastic diplegia: A basis for a management algorithm. *Eur. J. Neurol.* **2001**, *8* (Suppl. S5), 98–108. [CrossRef]
7. Jackman, M.; Sakzewski, L.; Morgan, C.; Boyd, R.N.; E Brennan, S.; Langdon, K.; Toovey, R.A.M.; Greaves, S.; Thorley, M.; Novak, I. Interventions to improve physical function for children and young people with cerebral palsy: International clinical practice guideline. *Dev. Med. Child Neurol.* **2022**, *64*, 536–549. [CrossRef]
8. Faccioli, S.; Pagliano, E.; Ferrari, A.; Maghini, C.; Siani, M.F.; Sgherri, G.; Cappetta, G.; Borelli, G.; Farella, G.M.; Foscan, M.; et al. Evidence-based management and motor rehabilitation of cerebral palsy children and adolescents: A systematic review. *Front. Neurol.* **2023**, *14*, 1171224. [CrossRef] [PubMed]
9. Booth, A.T.; Buizer, A.I.; Meyns, P.; Oude Lansink, I.L.; Steenbrink, F.; van der Krogt, M.M. The efficacy of functional gait training in children and young adults with cerebral palsy: A systematic review and meta-analysis. *Dev. Med. Child Neurol.* **2018**, *60*, 866–883. [CrossRef] [PubMed]
10. Hösl, M.; Schupfinger, A.; Klich, L.; Geest, L.; Bauer, P.; Bonfert, M.V.; Afifi, F.K.; Nader, S.; Berweck, S. Relationship between kinematic gait quality and caregiver-reported everyday mobility in children and youth with spastic Cerebral Palsy. *Eur. J. Paediatr. Neurol.* **2023**, *42*, 88–96. [CrossRef]

11. Park, E.Y. Path analysis of strength, spasticity, gross motor function, and health-related quality of life in children with spastic cerebral palsy. *Health Qual. Life Outcomes* **2018**, *16*, 70. [CrossRef]
12. Elnaggar, R.K.; Elbanna, M.F.; Mahmoud, W.S.; Alqahtani, B.A. Plyometric exercises: Subsequent changes of weight-bearing symmetry, muscle strength and walking performance in children with unilateral cerebral palsy. *J. Musculoskelet. Neuronal. Interact.* **2019**, *19*, 507–515. [PubMed]
13. Elnaggar, R.K. Acute responses to locomotor tasks differ according to gait-asymmetry patterns in children with hemiplegic cerebral palsy: An exploratory analysis. *Hum. Mov. Sci.* **2021**, *79*, 102860. [CrossRef] [PubMed]
14. Elnaggar, R.K.; Alhowimel, A.; Alotaibi, M.; Abdrabo, M.S.; Elshafey, M.A. Accommodating variable-resistance exercise enhance weight-bearing/gait symmetry and balance capability in children with hemiparetic cerebral palsy: A parallel-group, single-blinded randomized clinical trial. *Eur. J. Phys. Rehabil. Med.* **2022**, *58*, 378–386. [CrossRef]
15. Ibrahim, N.M.; Abbass, M.E. Weight distribution asymmetry in relation to walking speed in children with spastic cerebral palsy. *Afr. Health Sci.* **2022**, *22*, 565–572. [CrossRef]
16. Domagalska-Szopa, M.; Szopa, A.; Czamara, A. Dependence of gait deviation on weight-bearing asymmetry and postural instability in children with unilateral cerebral palsy. *PLoS ONE* **2016**, *11*, e0165583. [CrossRef]
17. Shumway-Cook, A.; Woollacott, M.H. *Motor Control: Translating Research into Clinical Practice*, 5th ed.; Lippincott Williams & Wilkins: Baltimore, MD, USA, 2016; pp. 660–754.
18. Elnaggar, R.K.; Alqahtani, B.A.; Alsubaie, S.F.; Mohamed, R.R.; Elbanna, M.F. Stretch-shortening cycle exercises can efficiently optimize gait-symmetry and balance capabilities in children with unilateral cerebral palsy: A randomized controlled trial. *NeuroRehabilitation* **2021**, *49*, 139–149. [CrossRef] [PubMed]
19. Elnaggar, R.K. An Analytical Insight into How Walking Speed and Spatial and Temporal Symmetry Are Related to Ankle Dysfunctions in Children with Hemiplegic Cerebral Palsy. *Am. J. Phys. Med. Rehabil.* **2021**, *100*, 458–464. [CrossRef]
20. Elnaggar, R.K. Relationship between Transverse-plane Kinematic Deviations of Lower Limbs and Gait Performance in Children with Unilateral Cerebral Palsy: A Descriptive Analysis. *Gait Posture* **2020**, *79*, 224–228. [CrossRef]
21. Grasso, R.; Zago, M.; Lacquaniti, F. Interactions between posture and locomotion: Motor patterns in humans walking with bent posture versus erect posture. *J. Neurophysiol.* **2000**, *83*, 288–300. [CrossRef]
22. Katz-Leurer, M.; Rotem, H.; Keren, O.; Meyer, S. Balance abilities and gait characteristics in post-traumatic brain injury, cerebral palsy and typically developed children. *Dev. Neurorehabil.* **2009**, *12*, 100–105. [CrossRef]
23. Chakraborty, S.; Nandy, A.; Kesar, T.M. Gait deficits and dynamic stability in children and adolescents with cerebral palsy: A systematic review and meta-analysis. *Clin. Biomech.* **2020**, *71*, 11–23. [CrossRef] [PubMed]
24. Patel, D.R.; Neelakantan, M.; Pandher, K.; Merrick, J. Cerebral palsy in children: A clinical overview. *Transl. Pediatr.* **2020**, *9* (Suppl. S1), S125–S135. [CrossRef] [PubMed]
25. Sutherland, D.; Olshen, R.; Biden, E.; Wyatt, M. *The Development of Mature Walking*; Cambridge University Press: Cambridge, UK, 1988.
26. Palisano, R.; Rosenbaum, P.; Walter, S.; Russell, D.; Wood, E.; Galuppi, B. Development and reliability of a system to classify gross motor function in children with cerebral palsy. *Dev. Med. Child Neurol.* **1997**, *39*, 214–223. [CrossRef]
27. Meseguer-Henarejos, A.B.; Sánchez-Meca, J.; López-Pina, J.A.; Carles-Hernández, R. Inter- and intra-rater reliability of the Modified Ashworth Scale: A systematic review and meta-analysis. *Eur. J. Phys. Rehabil. Med.* **2018**, *54*, 576–590. [CrossRef]
28. Davis, R.B.; Ounpuu, S.; Tyburski, D.; Gage, J.R. A gait analysis data collection and reduction technique. *Hum. Mov. Sci.* **1991**, *10*, 575–587. [CrossRef]
29. Biodex Medical Systems, Inc. *Balance System SD: Operation/Service Manual 950-441. Shirly*; Biodex Medical Systems, Inc.: New York, NY, USA, 2010.
30. An, C.M.; Son, Y.L.; Park, Y.H.; Moon, S.J. Relationship between dynamic balance and spatiotemporal gait symmetry in hemiplegic patients with chronic stroke. *Hong Kong Physiother. J.* **2017**, *37*, 19–24. [CrossRef] [PubMed]
31. Hickman, R.; Dufek, J.S.; Lee, S.P.; Blahovec, A.; Kuiken, A.; Riggins, H.; McClellan, J.R. Feasibility of using a large amplitude movement therapy to improve ambulatory function in children with cerebral palsy. *Physiother. Theory Pract.* **2015**, *31*, 382–389. [CrossRef]
32. Evkaya, A.; Karadag-Saygi, E.; Karali Bingul, D.; Giray, E. Validity and reliability of the Dynamic Gait Index in children with hemiplegic cerebral palsy. *Gait Posture* **2020**, *75*, 28–33. [CrossRef]
33. Zaino, C.A.; Marchese, V.G.; Westcott, S.L. Timed up and down stairs test: Preliminary reliability and validity of a new measure of functional mobility. *Pediatr. Phys. Ther.* **2004**, *16*, 90–98. [CrossRef]
34. Hsue, B.J.; Miller, F.; Su, F.C. The dynamic balance of the children with cerebral palsy and typical developing during gait. Part I: Spatial relationship between COM and COP trajectories. *Gait Posture* **2009**, *29*, 465–470. [CrossRef]
35. Patterson, K.K.; Mansfield, A.; Biasin, L.; Brunton, K.; Inness, E.L.; McIlroy, W.E. Longitudinal changes in poststroke spatiotemporal gait asymmetry over inpatient rehabilitation. *Neurorehabil. Neural. Repair* **2015**, *29*, 153–162. [CrossRef] [PubMed]
36. Perry, J.; Burnfield, J.M. Basic functions. In *Gait Analysis: Normal and Pathological Function*, 2nd ed.; Perry, J., Burnfield, J.M., Eds.; Slack Inc.: San Francisco, CA, USA, 2010; pp. 19–48.
37. Balasubramanian, C.K.; Bowden, M.G.; Neptune, R.R.; Kautz, S.A. Relationship between step length asymmetry and walking performance in subjects with chronic hemiparesis. *Arch. Phys. Med. Rehabil.* **2007**, *88*, 43–49. [CrossRef] [PubMed]

38. Lewek, M.D.; Bradley, C.E.; Wutzke, C.J.; Zinder, S.M. The relationship between spatiotemporal gait asymmetry and balance in individuals with chronic stroke. *J. Appl. Biomech.* **2014**, *30*, 31–36. [CrossRef] [PubMed]
39. Descatoire, A.; Femery, V.; Potdevin, F.; Moretto, P. Step-to-step reproducibility and asymmetry to study gait auto-optimization in healthy and cerebral palsied subjects. *Ann. Phys. Rehabil. Med.* **2009**, *52*, 319–329. [CrossRef]
40. Böhm, H.; Döderlein, L. Gait asymmetries in children with cerebral palsy: Do they deteriorate with running? *Gait Posture* **2012**, *35*, 322–327. [CrossRef]
41. Kenis-Coskun, O.; Giray, E.; Eren, B.; Ozkok, O.; Karadag-Saygi, E. Evaluation of postural stability in children with hemiplegic cerebral palsy. *J. Phys. Ther. Sci.* **2016**, *28*, 1398–1402. [CrossRef]

symmetry

MDPI

Article

Association of Core Muscle Endurance with Weekly Workout Time, Speed, and the Symmetry of Frontal Core Motion during Indoor Walking and Cycling

Ju-Hyun Park [1], Ae-Ryeong Kim [1], Si-Hyun Kim [2], Kwang Bok Kim [3] and Kyue-Nam Park [1,4,*]

1 Department of Rehabilitation Science, Jeonju University, Jeonju 55069, Korea
2 Department of Physical Therapy, Sangji University, Wonju 26339, Korea
3 Digital Health Care R&D Department, Korea Institute of Industrial Technology (KITECH), Cheonan 31056, Korea
4 Department of Physical Therapy, Jeonju University, Jeonju 55069, Korea
* Correspondence: knpark@jj.ac.kr; Tel.: +82-33-220-4664

Abstract: This study investigated the factors that influence core muscle endurance, i.e., the symmetry of frontal core motion during indoor walking and cycling, the symmetry of lateral core muscle endurance, the symmetry of the hip abductor strength, the weekly workout time and fast walking and cycling speeds, while controlling for gender. Seventy-nine healthy young adults participated in this study. In a regression analysis, the core muscle endurance time was the dependent variable. The independent variables were the symmetry of frontal core motion (measured using a wireless earbud sensor during walking and cycling), the symmetry of side plank time and of hip abductor strength, the weekly workout time and fast walking and cycling speeds. In the multiple regression analysis, weekly workout time, fast walking speed, symmetry of frontal core motion during fast cycling and symmetry of lateral side plank time predicted core muscle endurance (adjusted $R^2 = 0.42$). Thus, clinicians and fitness personnel should consider the association of core muscle endurance with the symmetry of frontal core motion during cycling and the symmetry of side plank holding time, as well as with the weekly workout time and a fast walking speed, when designing core muscle exercise programmes.

Keywords: core muscle endurance; cycling; home workout; symmetry; walking

1. Introduction

The body's core consists of the trunk, pelvic and hip regions, and these components form a kinetic chain with the extremities [1,2]. A better core performance results in a better sport performance, including a longer throwing distance and higher running speeds, as well as less injury, back pain and urinary incontinence [3–5]. Core performance has been evaluated using core muscle endurance tests, such as plank tests, to pre-screen for injuries of the lower extremities and to prescribe exercise programmes [6,7]. However, the measurement of the core muscle endurance time requires considerable effort and time, as well as a trained tester [8]. A model was developed to predict the endurance time in the performance of a side plank based on age, gender, body mass index and questionnaire-assessed variables [8]. There was a gender difference in lateral trunk endurance time, with males having greater holding times on both sides in side plank endurance tests than females [9]. However, there was no gender difference in trunk flexor endurance time [9].

Based on previous regression and correlation analyses, the weekly workout time is significantly associated with core muscle endurance; a longer weekly workout time had a positive correlation with better core muscle endurance [10]. Core muscle endurance is further related to the walking and indoor cycling speeds. Walking is the most popular physical activity worldwide [11], and indoor cycling was also very popular during the

COVID-19 pandemic worldwide [12]. Two studies showed that a core strengthening exercise programme can lead to improved cycling speed in cyclists and improved walking speed in patients with lumbar fusion, consistent with the association between core strength and speed performance [13,14].

The symmetry of core motion and lateral core muscle performance are the goals of rehabilitation during cycling and walking. Asymmetrical, increased lateral leaning of the trunk in the indoor cycling test is associated with a reduced activation of the core musculature, which increases the spinal load [15,16]. One study demonstrated asymmetrical pelvic tilt and hip rotation during indoor cycling in individuals with lower back pain and no leg length asymmetry, which was likely caused by a bilateral imbalance in the trunk muscles [15]. Another study revealed the asymmetrical activation of the lumbar multifidus in association with an asymmetrical lumbar rotation in the initial and final phases of indoor cycling in cyclists with lower back pain compared to healthy cyclists [17]. Asymmetrical lateral trunk leaning during walking has been also observed in people with knee osteoarthritis and patellofemoral pain [18]. With respect to the lateral core musculature, the weakness of the hip abductors may be related to asymmetrical lateral trunk leaning and pelvic drop during walking, resulting in the Trendelenburg gait [19]. In addition, asymmetrical lateral trunk leaning during walking affects the endurance of the lateral trunk muscles because their compensatory activation can be induced by weak hip abductors, in turn inducing their fatigue over time [20,21]. A prospective study also identified asymmetrical hip abduction strength as a potential risk factor for lower extremity injury [22].

Although symmetry is a desirable goal, its relationship to core muscle endurance during popular workouts, such as walking and cycling, or during performance tests, has yet to be investigated. Thus, this study included variables related to symmetry while performing workout activities, as well as previously identified variables, when investigating factors associated with core muscle endurance. The purpose of this study was to investigate factors that influence core muscle endurance, i.e., the symmetry of frontal core motion during indoor walking and cycling, the symmetry of lateral core muscle endurance, the symmetry of hip abductor strength, the weekly workout time, and indoor walking and cycling speeds, while controlling for gender. We hypothesised that core muscle endurance would have significant associations with the symmetry of core motion during cycling and walking, the hip strength, the lateral core muscle endurance bilaterally, the weekly workout time and the walking and cycling speeds.

2. Materials and Methods

2.1. Participants

This was a cross-sectional study with a convenience sample. The participants were recruited through online social media advertisements and via posters placed around a college campus and a local community in South Korea. The participants were interviewed by the researchers to confirm they met the inclusion criteria, which were as follows: age 19–30 years, body mass index <25 kg/m^2 and good health with no reported history of major physical discomfort or psychological symptoms that prevented their participation in the tests. This latter criterion was assessed by the following questions: "Do you suffer from any illness or injury of a physical or psychological nature that impairs your functioning in everyday life?"; "Have you ever had one of the following diseases or symptoms diagnosed by a doctor or self-reported: pregnancy, vertebral pathology (e.g., tumour, fracture or infection), cancer, lumbar surgery, psychiatric diagnosis, balance impairment related to dizziness, neurological disorder (e.g., spinal cord injury or central nervous system diseases), or chronic pain in the lower back and lower extremities for at least 3 months during the past year?"; and "Have you ever experienced discomfort while performing the plank or while walking or cycling quickly?" Only the participants who answered "no" to all questions were included in the study. All experimental procedures were explained prior to study participation, and written consent forms were obtained from all participants. This study was approved by the university's institutional review board (jjIRB-210114-HR-2021-0113).

2.2. Instruments

2.2.1. Strength Measurement System

A tensiometer (Smart KEMA pressure sensor; Factorial Holdings, Seoul, Korea) was used to measure the isometric strength of the hip abductor. The tensiometer measured 65 × 83 × 28 mm and weighed 110 g. To measure hip abductor strength, a 5 cm wide non-elastic strap was attached to the distal lower leg, with an absorber as a fixation point for firm attachment to the floor in the side-lying position. The sampling frequency was 10 Hz. The data were transferred to a Galaxy tablet (A6 10.1; Samsung, Seoul, Korea) via Bluetooth and analysed using Smart KEMA software (Factorial Holdings).

2.2.2. Wireless Earbud-Type IMU Sensor

Frontal core motion during walking and cycling was measured using a high-resolution inertial measurement unit (IMU; BNO080; CEVA Technologies, Rockville, MD., USA) consisting of an accelerometer and a gyroscope embedded into a wireless earbud (QCY-T6; Dongguan Hele Electronics Co., Ltd., Dongguan, China) that was worn on the participant's right side. The size and weight of the IMU sensor were 36 × 15 × 7.5 mm and 8.2 g, respectively. The orientation of the accelerometer was aligned with the gravitational axis corresponding to the standing position. The collected data were sent via Bluetooth to a computer. The sampling rate was fixed at 100 Hz. The recorded data were used to estimate the frontal core angle, calculated using Matlab (version R2018a; MathWorks, Natick, MA, USA). Prior to the calculation, the accelerometer output was filtered through a low-pass filter. Before the measurement was started, off calibration was performed automatically for 1 s. During the calibration period, 100 data samples were collected while the participant remained stationary.

2.3. Procedure

The experiments were conducted in a university laboratory in South Korea from February 2021 to January 2022. The experimental procedure consisted of four sessions that included the following: (1) baseline measurements and warm-up, (2) core muscle endurance test, (3) hip abductor strength test and (4) speed and frontal core motion tests during treadmill walking and indoor cycling. The experiment took about 80 min to complete, and each session took 20 min. The sequence of the experimental procedures after the warm-up was randomised using Excel (Microsoft Corp., Redmond, WA, USA), with a passive rest between sessions to minimise the fatigue effects arising due to repetitive exercising. The tests for strength, speed and frontal core motion were completed by one examiner, who was blinded to the core muscle endurance test.

2.3.1. Baseline Measurements and Warm-up

Baseline information on demographic characteristics (gender, height, weight, body mass index and weekly workout time) was collected. Each participant was asked to report the total weekly workout time including the number of days a week and the duration of each workout period [23]. Participants wore their own shoes and conducted a 5 min indoor cycling session for the warm-up, followed by 5 min of passive rest.

2.3.2. Core Muscle Endurance Test

The participants' core muscle endurance was assessed using the prone and side plank tests, with the order randomised to minimise the fatigue effects. The endurance times from the three plank tests (prone, left and right sides) were summed for data analysis [24]. In the prone plank test, the participants were prone, with their elbows in contact with the ground, such that the humerus was perpendicular to the horizontal plane, directly beneath the shoulders. The elbows were spaced shoulder-width apart, and the feet were close together. The participants were then instructed to raise the pelvis from the floor so that only the forearms and toes were in contact with the floor, while the shoulders, hips and ankles were maintained in a straight line (Figure 1a) [24]. In the side plank test, the participants lay on their sides with their legs extended. The upper foot was placed in front of the lower foot for support. Support was maintained using one elbow and one foot, and the hips were raised up from the floor, with the maintenance of a straight line along the lateral sides of the trunk and lower legs. The top arm was held across the chest with the hand placed on the opposite shoulder (Figure 1b) [21,25]. During the plank tests, the participants were asked to maintain the positions for as long as possible. The timer was stopped when the participant could no longer maintain a straight line between the trunk and the hip. Each plank test was performed once. To avoid muscle fatigue, a 5 min passive rest without any recovery exercise was allowed between each plank test.

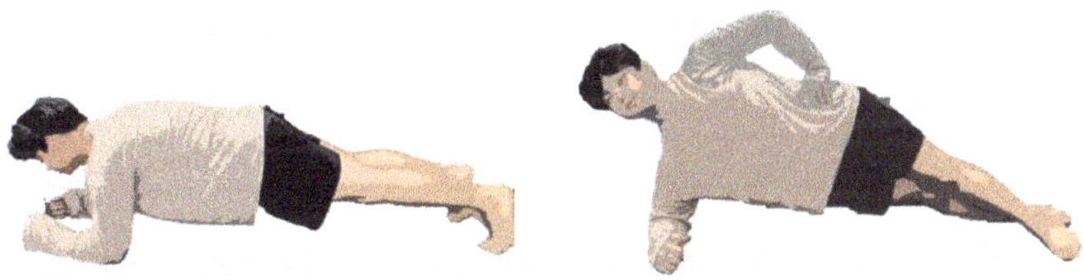

(a) (b)

Figure 1. Core muscle endurance tests: (**a**) prone plank test and (**b**) side plank test.

2.3.3. Hip Abductor Strength Test

The maximal isometric strength of the hip abductors was measured on each side using a tensiometer with a non-elastic band (smart KEMA pressure sensor; Factorial Holdings Co., Seoul, Korea), with participants lying on their sides. The pelvis was held to minimise compensatory pelvic elevation and rotation during the test. The participants were asked to extend the hip and knee on the tested side, with a 10° hip abduction. The hip and knee on the non-tested side were flexed slightly. The duration of the contractions was 5 s, and each contraction was repeated three times with a 30 s passive rest between repetitions. Both sides were tested, with a 5 min passive rest before switching the side (Figure 2a) [26].

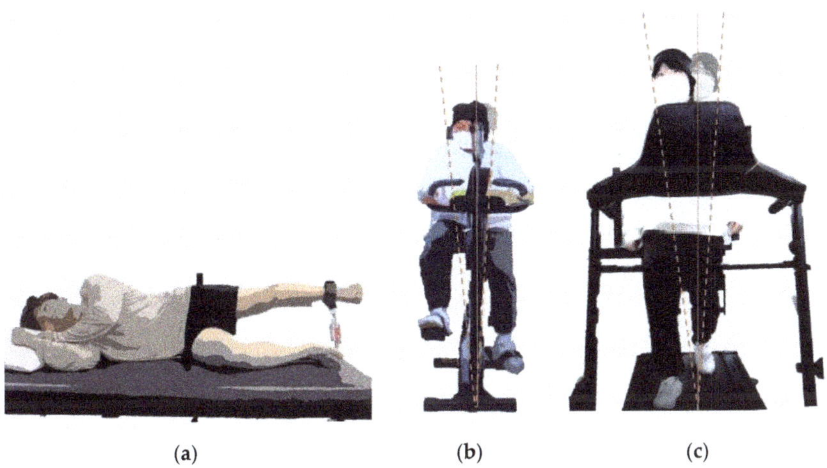

(a) (b) (c)

Figure 2. Symmetry tests: (**a**) symmetry of abductor strength in both hips; (**b**) symmetry of frontal core motion using a wireless earbud sensor during indoor cycling at fast speed; and (**c**) symmetry of frontal core motion using a wireless earbud sensor during treadmill walking at fast speed.

2.3.4. Speed and Frontal Core Motion Tests during Treadmill Walking and Indoor Cycling

A wireless earbud-type IMU sensor was used in these tests because the wireless earbud is a popular wearable device that is commonly used to listen to music during workouts and because it is worn closer to the trunk than other popular wearable devices, such as smartwatches [27]. The assessment of core motion in the frontal plane using an earbud-type IMU sensor has been shown to be valid compared with the use of a 3D motion analysis system, and the data correlate with trunk motion during home workout activities [27]. Frontal core motion data were collected during 60 s of walking or cycling.

Indoor Cycling at a Fast Speed

The participants cycled at a self-selected fast speed on an indoor stationary bicycle. Each participant adjusted the cycle seat height to a comfortable position. Before data collection, the participants were allowed 3 min to become familiar with the self-selected fast speed. They were instructed to 'cycle as fast as possible' [28]. After a 60 s passive rest, the participants cycled for 1 min. During cycling, they held the front handlebars with both hands and fixed their gazes on the cycle's speedometer (Figure 2b).

Treadmill Walking at a Fast Speed

The participants walked for 3 min on a treadmill to determine their self-selected fast walking speed. The verbal instructions that they received were 'walk as fast as possible without running'. The participants switched the treadmill on and then gradually increased the speed by 0.5 km/h until their fast walking speed was selected [28]. After 60 s of passive rest, the participants walked on the treadmill for 1 min. While walking, they fixed their gaze on the front tablet at eye level (Figure 2c).

2.4. *Data Analysis*

Regarding the frontal core motion data, the middle 50 s of the 60 s walk or cycle were analysed, excluding the 5 s initial and final deceleration phases. A symmetry index (SI) was used to quantify the asymmetry of the right and left sides for the side plank test, hip abduction strength and frontal core motion during walking and cycling, using the formula [29]:

$$\text{Symmety Index} = 100 - \left| 100 \times \frac{Right\ side - Left\ side}{Right\ side + Left\ side} \right|$$

The SI was expressed as a percentage, with a value of 100 indicating absolute symmetry, and lower values indicating greater asymmetry between the right and the left sides.

2.5. *Statistical Analysis*

All variables were summarised with standard descriptive statistics, including the mean and standard deviation (SD). Normality was confirmed using the Shapiro–Wilk test. Independent t-tests or the Mann–Whitney U test were used to compare the variables between genders. Univariate regression analysis was used to examine the relationships between core muscle endurance time and each independent variable. i.e., gender, weekly workout time, fast walking and cycling speeds, SIs for hip abduction strength, side plank endurance time and frontal core motion during fast walking and cycling. A multiple regression analysis using stepwise selection was performed to determine which of the following factors had the greatest influence on core muscle endurance time after adjusting for gender: weekly workout time, fast walking and cycling speeds and SIs for hip abduction strength, SIs for side plank endurance time and SIs for frontal core motion during fast walking and cycling. All models were adjusted for gender. SPSS (version 25.0; IBM Corp., Armonk, NY, USA) was used for the statistical analyses. The alpha level was set to 0.05.

3. Results

This study enrolled 79 healthy young adults (31 males and 48 females, mean ± SD age, 21.80 ± 2.78 years; body mass index, 21.42 ± 2.11 kg/m^2). Table 1 presents the participants' characteristics and compares the variables between genders. The mean and SD for all variables were significantly greater for males than for females ($p < 0.05$), except the SI for frontal core motion during fast walking ($p = 0.18$). Table 2 provides the descriptive data for the right and left sides obtained before calculating the SIs using the formula.

Table 1. Descriptive statistics for variables by gender and overall.

Variables	Males (n = 31)	Females (n = 48)	Total (n = 79)	p Value
Core muscle endurance time, s	260.81 ± 63.29	185.99 ± 81.99	215.35 ± 83.33	0.01
Weekly workout time, min	365.32 ± 195.48	235.46 ± 264.56	286.40 ± 246.87	0.01
Fast walking speed, km/h	6.42 ± 1.20	5.24 ± 1.02	5.70 ± 1.23	0.01
Fast cycling speed, km/h	39.87 ± 3.87	35.61 ± 4.32	37.28 ± 4.63	0.01
SI * for the side plank endurance time, %	91.10 ± 8.48	86.02 ± 12.07	88.01 ± 11.03	0.04
SI for hip abduction strength, %	94.49 ± 3.55	91.65 ± 5.83	92.76 ± 5.23	0.04
SI for frontal core motion during fast walking speed, %	79.18 ± 18.40	70.25 ± 24.52	73.75 ± 22.62	0.18
SI for frontal core motion during fast cycling speed, %	75.71 ± 19.44	59.89 ± 30.13	66.10 ± 27.44	0.01

* SI, symmetry index. Values are presented as mean ± SD.

Table 2. Descriptive data on both sides for the variables that were calculated using the symmetry index.

Variables	Mean ± SD
Side plank endurance time (left/right), s	65.33 ± 31.50/64.14 ± 28.00
Hip abduction strength (left/right), kgf	11.44 ± 4.45/11.67 ± 4.50
Frontal core motion during fast walking speed (left/right), °	5.89 ± 3.30/5.91 ± 2.34
Frontal core motion during fast cycling speed (left/right), °	12.30 ± 18.30/12.51 ± 11.79

Table 3 presents the results of the univariate linear regression analyses. Gender was significantly associated with the core muscle endurance time. All other variables, except SI for frontal core motion during fast walking, were also significantly associated with the core muscle endurance time (Table 3).

Table 3. Results of univariate regression analyses with core muscle endurance.

Independent Variable	ß Coefficient *	95% CI	p
Gender	74.82	40.29, 109.35	0.00
Weekly workout time, min	0.16	0.09, 0.23	0.00
Fast walking speed, km/h	31.48	17.89, 45.07	0.00
Fast cycling speed, km/h	8.61	5.02, 12.20	0.00
SI # for hip abduction strength, %	4.96	1.52, 8.40	0.01
SI for side plank endurance time, %	2.92	1.34, 4.50	0.00
SI for frontal core motion during fast walking speed, %	0.62	−0.02, 1.45	0.14
SI for frontal core motion during fast cycling speed, %	0.94	0.29, 1.60	0.01

* ß coefficient represents the estimated change in seconds in core muscle endurance time for 1 unit. change in the independent variable. # SI, symmetry index.

Five models were built from the multiple stepwise regression analysis after adjusting for gender. The final model explained 42% of the variance in core muscle endurance and included four variables: weekly workout time, SI for the side plank endurance time, fast walking speed and SI for frontal core motion during fast cycling (Table 4). Three non-significant variables were not included in the final model: fast cycling speed (ß Coefficient = 0.14, $p = 0.18$), the SI for hip abduction strength (ß Coefficient = 0.14, $p = 0.14$) and the SI for frontal core motion during fast walking speed (ß Coefficient = 0.01, $p = 0.94$).

Table 4. Results of stepwise multivariate regression analyses with core muscle endurance, adjusted for gender.

Selected Variables in the Final Model	R^2	ΔR^2	Standardized ß *	t	*p*
Gender			0.16	1.54	0.13
Weekly workout time, min			0.33	3.61	0.00
SI # for the side plank endurance time, %	0.46	0.42	0.20	2.19	0.03
Fast walking speed, km/h			0.22	2.11	0.04
SI for frontal core motion during fast cycling speed, %			0.19	2.06	0.04

* Standardized ß coefficient represents the magnitude of the contribution that each predictor variable makes to maximally predicting the core muscle endurance time in the regression model. # SI, symmetry index.

4. Discussion

This study examined the associations between factors related to a home workout setting and core muscle endurance and identified predictors of core muscle endurance with adjustment for gender. Males had greater core muscle endurance, longer weekly workout times, higher speeds, higher SIs for side plank endurance time and greater hip abduction strength and frontal core motion during fast cycling than females (Table 1). A previous study also demonstrated gender differences; specifically, males had better hip muscle strength and core muscle endurance times [25]. In addition, univariate analyses showed significant relationships between gender and all variables, except for the symmetry of core motion during fast walking (Table 3). However, this gender difference did not remain after adjusting for other covariates, indicating the major confounder of the gender–core muscle endurance relationship (Table 4). In line with the current study, gender was a potential confounder in a previous multiple regression analysis of factors influencing back muscle endurance [10].

In the current study, symmetry of frontal core motion during fast cycling, symmetry of the side plank endurance time, fast walking speed and weekly workout time accounted for 42% of the variance in core muscular endurance (Table 4). In a previous regression model, perceived self-efficacy, sitting trunk angle, weekly workout time and duration of daily TV use accounted for 15% of the variance in back muscle endurance [10]. In another study, body mass index, fat mass and body fat percentage accounted for 29–37% of the variance in core muscular endurance [30]. Our study included variables related to the symmetry of core motion, symmetry of lateral core endurance and walking speed, which were not considered as independent variables in the previous regression analysis. This increased the power to predict core muscle endurance relative to previous studies.

The symmetry of frontal core motion during fast cycling (mean ± SD, 66.10 ± 27.44%) contributed to core muscle endurance (mean ± SD, 215.35 ± 83.33 s) (Tables 1 and 4). Among the cyclists (8 males and 10 females) participating in the previous study, asymmetrical lower lumbar rotation was observed at the initial and final phases of indoor cycling in those with lower back pain compared with healthy cyclists [17]. Asymmetrical activation of the multifidus in the lower lumbar region was also observed during indoor cycling in cyclists with lower back pain [17]. Asymmetrical trunk motion in the frontal plane might have occurred in association with asymmetrical core muscle activation in our study. Along with asymmetrical trunk motion and core muscle activation, asymmetrical pedalling was observed in non-professional and professional cyclists with low core stability and body asymmetry, as measured using a functional movement screening test [31,32]. The above findings, together with those of the current study, indicate that core muscle endurance can be predicted from the asymmetry of the trunk in the frontal plane during fast cycling.

The final regression model included the SI for side plank endurance time (Table 4). This result means that the side plank symmetry contributed significantly to the core muscle endurance time, in contrast to the findings of a previous study in which the symmetry of the side plank endurance time was not determined [25]. The difference between these studies is that the body was supported by the elbow and one foot in the latter [25] and by the elbow and both feet, with the top foot placed in front of the lower foot to provide additional support, in our study. In addition, the participants in the previous study were healthy, physically active Navy cadets [25]. The weekly workout times of our participants ranged from 0 to 1050 min (mean $\pm$ SD, 286.40 $\pm$ 246.87 min) and contributed significantly to the core muscle endurance time (Tables 1 and 4). Previous regression and correlation analyses support our finding that the weekly workout time is associated with core muscle endurance; specifically, a longer weekly workout time had a positive correlation with core muscle endurance [10]. Therefore, the symmetry of the side plank endurance time along with the weekly workout time should be considered when assessing the core performance.

In a multiple regression analysis, fast walking speed was identified as a factor influencing the core muscle endurance (Table 4). A previous study demonstrated that gradually increasing the walking speed to 2, 3, 4, 5 and 6 km/h induced a greater activation of the core muscles (the rectus abdominis and internal and external obliques), indicating that greater core muscle contraction is necessary when the walking speed is increased [33]. The results of this and previous studies suggest that fast walking can be achieved with the improvement of core performance [13,33]. Clinicians and sports trainers should thus recommend fast walking to improve core muscle endurance.

The limitations of this study must be taken into account. First, the participants were healthy and young. Whether the results of this study also apply to the core muscle endurance of participants of different ages or with musculoskeletal disorders is unclear. Second, both males and females were included in the current study, and there was a gender difference in core muscle endurance time, among other variables. Although the regression analysis was adjusted for gender, a future study should investigate the factors contributing to core muscle endurance in each gender. Lastly, the study's cross-sectional design hindered the determination of whether an improvement in walking speed or the symmetry of frontal core motion during cycling improves core muscle endurance.

5. Conclusions

This study showed that core muscle endurance times were related to fast walking speed, weekly workout time, symmetry of frontal core motion during fast cycling and symmetry of side plank endurance times. Given that symmetry, speed and weekly workout time are associated with the core muscle endurance time, healthcare providers should adequately evaluate the symmetry of frontal core motion during cycling and side plank endurance in individuals who want to improve their core muscle endurance, as well as the weekly workout time and fast walking speed.

Author Contributions: Conceptualisation, S.-H.K. and K.-N.P.; data curation, K.B.K.; formal analysis, A.-R.K., J.-H.P. and K.B.K.; funding acquisition, K.-N.P.; methodology, A.-R.K., J.-H.P. and K.-N.P.; supervision, K.-N.P.; writing—original draft preparation, J.-H.P.; writing—review and editing, S.-H.K., K.B.K. and K.-N.P. All authors have read and agreed to the published version of the manuscript.

Funding: This work was supported by a National Research Foundation of Korea (NRF) grant funded by the Korean government (MSIT) (No. 2020R1A2C2102729).

Institutional Review Board Statement: The study was conducted according to the guidelines of the Declaration of Helsinki and approved by the Institutional Review Board of Jeonju University (jjIRB-210114-HR-2021-0113).

Informed Consent Statement: Informed consent was obtained from all participants involved in the study.

Data Availability Statement: The data presented in this study are available on request from the corresponding author, as they contain data from human volunteers.

Conflicts of Interest: The authors declare no conflict of interest.

References

1. Bird, M.; Fletcher, K.M.; Koch, A.J. Electromyographic comparison of the ab-slide and crunch exercises. *J. Strength Cond. Res.* **2006**, *20*, 436–440. [CrossRef] [PubMed]
2. Kibler, W.B.; Press, J.; Sciascia, A. The role of core stability in athletic function. *Sports Med.* **2006**, *36*, 189–198. [CrossRef]
3. Zazulak, B.T.; Hewett, T.E.; Reeves, N.P.; Goldberg, B.; Cholewicki, J. Deficits in neuromuscular control of the trunk predict knee injury risk: A prospective biomechanical-epidemiologic study. *Am. J. Sports Med.* **2007**, *35*, 1123–1130. [CrossRef] [PubMed]
4. Tong, T.K.; Wu, S.; Nie, J. Sport-specific endurance plank test for evaluation of global core muscle function. *Phys. Ther. Sport.* **2014**, *15*, 58–63. [CrossRef]
5. Smith, M.D.; Coppieters, M.W.; Hodges, P.W. Postural response of the pelvic floor and abdominal muscles in women with and without incontinence. *Neurourol. Urodyn.* **2007**, *26*, 377–385. [CrossRef]
6. Strand, S.L.; Hjelm, J.; Shoepe, T.C.; Fajardo, M.A. Norms for an isometric muscle endurance test. *J. Hum. Kinet.* **2014**, *40*, 93–102. [CrossRef] [PubMed]
7. Sciascia, A.; Cromwell, R. Kinetic chain rehabilitation: A theoretical framework. *Rehabil. Res. Pract.* **2012**, *2012*, 853037. [CrossRef]
8. Akay, M.F.; Yüksel, M.C.; Abut, F.; Taş, F.M.; George, J. Predicting the maximum endurance time for left-side bridge exercise using machine learning methods and hybrid data. In Proceedings of the 2017 9th International Conference on Computational Intelligence and Communication Networks (CICN), Girne, Northern Cyprus, 16–17 September 2017. [CrossRef]
9. Evans, K.; Refshauge, K.M.; Adams, R. Trunk muscle endurance tests: Reliability, and gender differences in athletes. *J. Sci. Med. Sport* **2007**, *10*, 447–455. [CrossRef]
10. Smith, A.J.; O'Sullivan, P.B.; Campbell, A.; Straker, L. The relationship between back muscle endurance and physical, lifestyle, and psychological factors in adolescents. *J. Orthop. Sports Phys. Ther.* **2010**, *40*, 517–523. [CrossRef]
11. Cesare, N.; Nguyen, Q.C.; Grant, C.; Nsoesie, E.O. Social media captures demographic and regional physical activity. *BMJ Open Sport Exerc. Med.* **2019**, *5*, e000567. [CrossRef]
12. Zirgaitis, G.; Beilfuss, R. Physical Activity Patterns and the Way the World Moves Alters during Stay at Home Orders. Available online: https://www.marquette.edu/innovation/documents/commentary-zirgaitis.pdf (accessed on 7 September 2020).
13. Wiseman, K. An Investigation into the Effectiveness of Core Muscle Strengthening on Cycling Performance in Asymptomatic Cyclists. Available online: http://hdl.handle.net/10321/963 (accessed on 8 April 2014).
14. Kernc, D.; Strojnik, V.; Vengust, R. Early initiation of a strength training based rehabilitation after lumbar spine fusion improves core muscle strength: A randomized controlled trial. *J. Orthop. Surg. Res.* **2018**, *13*, 151. [CrossRef]
15. Joseph, S.; Ganason, R.; Jalil, A.P.; Aizam, Z.S.; Wilson, B. A functional biomechanical analysis of an elite cyclist. In Proceedings of the ISBS-Conference Proceedings Archive, Salzburg, Austria, 14–18 July 2006. [CrossRef]
16. Galindo-Martínez, A.; López-Valenciano, A.; Albaladejo-García, C.; Vallés-González, J.M.; Elvira, J.L.L. Changes in the trunk and lower extremity kinematics due to fatigue can predispose to chronic injuries in cycling. *Int. J. Environ. Res. Public Health* **2021**, *18*, 3719. [CrossRef]
17. Burnett, A.F.; Cornelius, M.W.; Dankaerts, W.; O'Sullivan, P.B. Spinal kinematics and trunk muscle activity in cyclists: A comparison between healthy controls and non-specific chronic low back pain subjects-a pilot investigation. *Man. Ther.* **2004**, *9*, 211–219. [CrossRef] [PubMed]
18. van der Esch, M.; Steultjens, M.P.; Harlaar, J.; van den Noort, J.C.; Knol, D.L.; Dekker, J. Lateral trunk motion and knee pain in osteoarthritis of the knee: A cross-sectional study. *BMC Musculoskelet. Disord.* **2011**, *12*, 141. [CrossRef]
19. Takacs, J.; Hunt, M.A. The effect of contralateral pelvic drop and trunk lean on frontal plane knee biomechanics during single limb standing. *J. Biomech.* **2012**, *45*, 2791–2796. [CrossRef] [PubMed]
20. Seidenberg, P.H.; Bowen, J.D.; King, D.J. *The Hip and Pelvis in Sports Medicine and Primary Care*; Springer: New York, NY, USA, 2010.
21. Elataar, F.F.; Abdelmajeed, S.F.; Abdellatif, N.M.N.; Mohammed, M.M. Core muscles' endurance in flexible flatfeet: A cross-sectional study. *J. Musculoskelet. Neuronal. Interact.* **2020**, *20*, 404–410.
22. Blaiser, C.D.; Roosen, P.; Willems, T.; Bleecker, C.D.; Vermeulen, S.; Danneels, L.; Ridder, R.D. The role of core stability in the development of non-contact acute lower extremity injuries in an athletic population: A prospective study. *Phys. Ther. Sport* **2021**, *47*, 165–172. [CrossRef]
23. Chou, Y.J.; Lai, Y.H.; Lin, B.R.; Liang, J.T.; Shun, S.C. Factors influencing amount of weekly exercise time in colorectal cancer survivors. *Cancer. Nurs.* **2017**, *40*, 201–208. [CrossRef] [PubMed]
24. Imai, A.; Kaneoka, K. The relationship between trunk endurance plank tests and athletic performance tests in adolescent soccer players. *Int. J. Sports Phys. Ther.* **2016**, *11*, 718–724.
25. Lopes, T.J.A.; Simic, M.; Alves, D.S.; Bunn, P.D.S.; Rodrigues, A.I.; Terra, B.S.; Lima, M.D.S.; Ribeiro, F.M.; Vilão, P.; Pappas, E. Physical performance measures of flexibility, hip strength, lower limb power, and trunk endurance in healthy navy cadets: Normative data and differences between sex and limb dominance. *J. Strength Cond. Res.* **2021**, *35*, 458–464. [CrossRef]

26. Kim, S.H.; Jeong, S.W.; Park, K.N. Classification model to discriminate people with and without pain in the lower back and lower limb using symmetry data. *J. Musculoskelet. Sci. Technol.* **2021**, *5*, 72–79. [CrossRef]
27. Kim, A.R.; Park, J.H.; Kim, S.H.; Kim, K.B.; Park, K.N. The validity of wireless earbud-type wearable sensors for head angle estimation and the relationships of head with trunk, pelvis, hip, and knee during workouts. *Sensors* **2022**, *22*, 597. [CrossRef] [PubMed]
28. Damiano, D.L.; Norman, T.; Stanley, C.J.; Park, H.S. Comparison of elliptical training, stationary cycling, treadmill walking and overground walking. *Gait. Posture* **2011**, *34*, 260–264. [CrossRef]
29. Agrawal, V.; Gailey, R.; O'Toole, C.; Gaunaurd, I.; Dowell, T. Symmetry in external work (SEW): A novel method of quantifying gait differences between prosthetic feet. *Prosthet. Orthot. Int.* **2009**, *33*, 148–156. [CrossRef]
30. Mayer, J.M.; Nuzzo, J.L.; Chen, R.; Quillen, W.S.; Verna, J.L.; Miro, R.; Dagenais, S. The impact of obesity on back and core muscular endurance in firefighters. *J. Obes.* **2012**, *2012*, 729283. [CrossRef] [PubMed]
31. Frohm, A.; Heijne, A.; Kowalski, J.; Svensson, P.; Myklebust, G. A nine-test screening battery for athletes: A reliability study. *Scand J. Med. Sci. Sports* **2012**, *22*, 306–315. [CrossRef] [PubMed]
32. Rannama, I.; Pedak, K.; Reinpõld, K.; Port, K. Pedalling technique and postural stability during incremental cycling exercise–relationship with cyclist FMS TM score. *LASE J. Sport Sci.* **2016**, *7*, 1–18. [CrossRef]
33. Anders, C.; Wagner, H.; Puta, C.; Grassme, R.; Petrovitch, A.; Scholle, H.C. Trunk muscle activation patterns during walking at different speeds. *J. Electromyogr. Kinesiol.* **2007**, *17*, 245–252. [CrossRef] [PubMed]

symmetry

MDPI

Article

Is Asymmetry Different Depending on How It Is Calculated?

Mario Iglesias-Caamaño, Tania Álvarez-Yates , Javier Carballo-López, Alba Cuba-Dorado and Óscar García-García *

Laboratory of Sports Performance, Physical Condition and Wellness, Faculty of Education and Sport Sciences, University of Vigo, 36310 Pontevedra, Spain

* Correspondence: oscargarcia@uvigo.es; Tel.: +34-986-801-782

Abstract: This study aimed to (1) determine the magnitude and direction of asymmetry in volleyball players, (2) establish asymmetry thresholds, and (3) explore differences depending on the test used and the players' category. Twenty-nine junior and senior male volleyball players were assessed through a muscle asymmetry battery test: active knee extension test (AKE), single-leg countermovement jump (SL-CMJ), single-leg squat jump (SL-SJ), triple hop test for distance (THTD), modified 20-yard shuttle run, Y-balance test, single-leg one-repetition maximum in leg press test (1RM-SL), and lateral symmetry in radial muscle belly displacement through Tensiomyography in the biceps femoris and rectus femoris. A two-way ANOVA alongside an individual analysis of asymmetry thresholds was used to analyze the test and categorize the influence on the magnitude and the direction of asymmetry. The 1RM-SL, SL-SJ, and the lateral symmetry in radial muscle belly displacement showed a clear asymmetry towards the non-dominant side, while the AKE, SL-CMJ, and THTD showed an asymmetry towards the dominant side. The magnitude of the asymmetry was highly variable between tests (1.46–30.26%). The individualized asymmetry thresholds revealed that the percentage of asymmetrical players varied depending on the type of test used. In conclusion, the type of test used determines the magnitude and direction of asymmetry in well-trained volleyball players.

Keywords: asymmetry thresholds; inter-limb differences; imbalance; volleyball

1. Introduction

Volleyball is a sport characterized by high-intensity short duration and explosive actions with interspersed rest phases [1]. The dynamics of the game involve multiple actions such as jumps, landings, sprints, and multi-directional movements, which emphasize the importance of volleyball players' neuromuscular system. The jump is one of the most outstanding actions in the matches, which differs depending on a player's position. Setters can perform about 136 jumps per game, while middles, opposites and hitters around 97, 88 and 65 jumps respectively [2]. Therefore, it has been suggested that these athletes require well-developed muscle power and speed [3].

Lateral dominance may result in asymmetries, which can be understood as differences in strength, power, stiffness, or range of motion (ROM) between both sides of the player's body [4]. The typical high-intensity actions of team sports (i.e., changes of direction, jumps, hitting, throwing) tend to have a one-sided appearance [5] which can lead to sport-specific adaptions that may entail overloads of body structures [6] and a markedly asymmetric force generations in lower limbs [7]. In volleyball, the jump take-off is performed with one leg ahead depending on the volleyball player's laterality, while the landing is primarily performed one-legged in certain specific varieties of the attack [8–10]. In addition, there are specific zones gestures such as blocks that are normally performed towards the same side [11]. Thus, these repetitive actions in training and competitions can favor the development of asymmetries in volleyball.

Lateral asymmetries have been studied in the scientific literature according to the characteristics of the limbs. Some authors have referred to each leg as the dominant (DL)

or non-dominant leg (NDL) [7], the strong or weak leg [12], or just the right and left leg [13]. Yet, all report the inter-limb percentage difference. However, there does not seem to be an absolute consensus on how much a greater or lesser asymmetry affects a player's performance. It has been pointed out that less asymmetry between extremities can improve specific performance actions such as changes of direction [14] or jumping performance [15]. However, other authors have not found significant positive relations between having less asymmetry and better performance [16]. In addition, the asymmetry of volleyball players based on age has not yet been addressed. However, authors such as Loturco et al. [17] have pointed out that the specific adaptations after years of sport specialization in high-level athletes can create inter-limb imbalances.

In the last few years, a bilateral difference was considered as an asymmetry when its magnitude exceeded 10–15% [12,18–20], based on the fact that this difference was associated with a higher risk of injury [21]. However, no consensus has yet been reached on the magnitude of the asymmetry, since this percentage is contemplated as an "arbitrary threshold" which cannot be applied to all assessments nor all populations [22]. Therefore, different procedures have been suggested to determine this percentage of asymmetry in athletes [23]. On the one hand, using the formula mean + (0.2 × SD) [23], was able to classify 42% of athletes with a "small to moderate" inter-limb asymmetry, while on the other hand, applying the more conservative formula mean + (1.0 × DE between subjects), only 16% of athletes were classified as asymmetrical, but with "high or extreme" limb asymmetry. In the first formula, it has been suggested that multiplying 0.2 × SD in elite team sports athletes produces the smallest worthwhile change [24] based on Cohen's effect size. Nevertheless, asymmetry should be individualized [16], as it rarely favors the same limb in all tests [5,25], so it will allow coaches to use these thresholds as reference, criteria, and normative data for specific populations and a variety of metrics and tests. However, to be significant, it needs to be verified that its value of asymmetry (%) must be greater than the coefficient of variation (CV) of the test [26].

The magnitude and direction of asymmetry in well-trained or professional volleyball players has been analyzed very little and has shown inconclusive findings. From a technical point of view, the attack hit has been considered as an asymmetrical action [9], since the forward leg supports most of the load [27]. However, no contralateral differences in the lower limbs were found after assessing asymmetry through a bilateral isokinetic test in German third league players [27] nor with a countermovement jump test (CMJ) in the first Brazilian league [28]. Nevertheless, other authors have observed lower limb asymmetries greater than 10% through a single-leg CMJ (SL-CMJ) test in players with 6 to 10 h training/week [29]. Yet, it is noted that "due to the multi-factorial nature of jumping performance, individual parameters related to performance may not be consistently different" [30]. In this sense, inter-limb differences are suggested to exist in volleyball players; however, the direction of asymmetry appears highly variable depending on the test used to evaluate the athlete [5].

These findings are still inconclusive, mainly due to the different tests that have been performed, the performance level of the players, and the reference percentage to indicate asymmetry. Therefore, it is essential to investigate the magnitude and direction of asymmetry and the reference asymmetry thresholds for well-trained volleyball players to correctly classify players with asymmetry issues. Likewise, it is crucial to establish the differences in asymmetry depending on the test used and the player's performance level (juniors vs. seniors).

Thus, the aims of this study were to determine the magnitude and direction of asymmetry in well-trained volleyball players and to establish asymmetry thresholds through a specific asymmetry battery test for volleyball. Additionally, this study aims to determine the differences in the magnitude and direction of asymmetry, as for the asymmetry thresholds, depending on the tests used and the volleyball players' category (juniors vs. seniors).

2. Materials and Methods

2.1. Study Design

A comparative and cross-sectional study design has been used, following an associative strategy, to determine the direction and magnitude of lateral asymmetry in well-trained volleyball players. The testing battery designed by Iglesias-Caamaño et al. [4,31] to assess muscle asymmetry in volleyball was performed within the players' competitive period.

To determine the magnitude and the direction of asymmetry, the bilateral strength asymmetry (BSA) formula modified by Bishop et al. [5] was used: $([DL - NDL)/DL \times 100] \times IF(DL < NDL, 1, -1)$. This formula implements the excel IF function that allows the monitoring of the direction of asymmetry without magnitude variation issues (expressed in %). For this purpose, the attack jump (AJ) leg was chosen as the DL (the left leg for all participants).

2.2. Participants

The sample was composed of 29 male volleyball players from the Spanish Superleague 2 (11 juniors and 18 seniors from different clubs) with a minimum of 8 h training/week and 9 years of volleyball practice experience. The players had an average age of 20.4 ± 4.2 years and an average height of 181 ± 8.0 cm (Stadiometer Seca 213, Seca gmbh and co. kg., Hamburg, Germany). Players' body composition was analyzed through a bioelectrical impedance scale (Tanita BC-601 Segment, Tanita Corporation, Tokyo, Japan), showing an average 75.3 ± 9.1 kg body mass, 12.4 ± 5.2% body fat, and 67.9 ± 9.8 kg muscle mass (see Table 1). All players who volunteered to participate in the study signed an informed consent form before data collection. Coaches and clubs' directors were also informed and approved the testing protocol. The research followed the ethical principles of the Declaration of Helsinki (64 Ed. 2013) and was approved by the Ethics Research Committee of the Faculty of Education and Sports Sciences of the University of Vigo (06-1019).

Table 1. Volleyball players' characteristics based on category.

Category	N	Age (Years)	Height (cm)	Body Mass (kg)	Body Fat (%)	Muscle Mass (kg)
Junior	11	16.5 ± 0.52	179.6 ± 8.6	73.0 ± 10.1	10.6 ± 4.1	61.8 ± 8.3
Senior	18	22.8 ± 3.6	182.1 ± 7.7	76.8 ± 8.5	13.6 ±5.6	71.7 ± 9.0

2.3. Procedures

Before starting the testing protocol, all players carried out their usual warmup. First, they performed 5 min of general mobility, followed by several body weight exercises, and then ended with a free simulation of each test 3 min prior testing. The players made three attempts in each test, retaining the best performance value for analysis [4].

2.3.1. Active Knee Extension (AKE)

The AKE test was used to assess players' hamstring flexibility symmetry. This test has previously been described by Iglesias-Caamaño et al. [4] and is based on angular measurements. A digital goniometer (Baseline Absolute Axis 360°, Fabrication Enterprises, Inc., White Plains, NY, USA) was used to measure players' knee extension with the hip fixed at 90° [32].

2.3.2. Single-Leg Countermovement Jump (SL-CMJ)

The SL-CMJ test was performed on the Chronojump contact platform (Chronojump Bosco-System®, Barcelona, Spain) following the protocol described by Thomas et al. [33]. Players' preformed a single-leg jump with a free flexion angle, but during the entire jump, their hands had to be fixed on the hips. Flight time (s) was retained for data analysis, which was provided by the Chronojump software platform (v. 1.7.0 for Windows) as a direct measure.

2.3.3. Modified 20-Yard Shuttle Run (M-20Y)

The M-20Y test assessed the time spent by the players to cover 20 m with two turns of 180°. The protocol of Sekulic et al. [34] was followed, but subtly modified to perform it unilaterally, varying the turning leg in each attempt as previously described by Iglesias-Camaaño et al. [4] (Figure 1). Microgate polifemo photocells (Microgate Corporation, Bolzano, Italy, Software version 1.10.19.01) were used to measure the time to cover the 20 m distance. Each player performed 6 attempts, 3 turns on the right leg (RL) and 3 turns on the left leg (LL), with 1 min break.

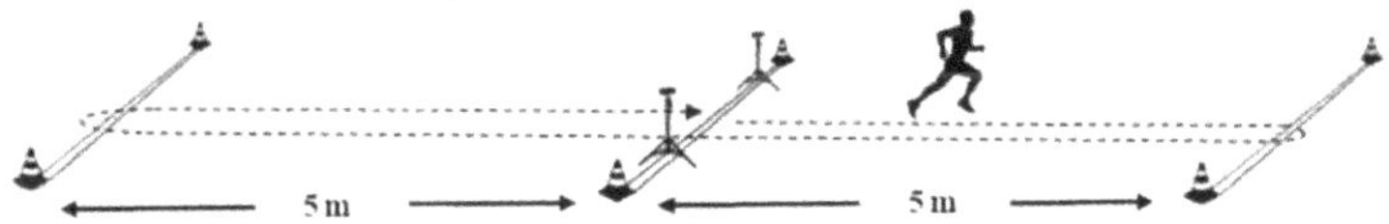

Figure 1. Modified 20-yard shuttle run.

2.3.4. Y-Balance Test (YBT)

The YBT was used to assess the players' lower limbs dynamic balance symmetry using the Y-Balance Test Kit™ (Functionalmovement.com, Danville, VA, USA). Prior to the test, players' lower limb lengths (cm) were measured from the anterior superior iliac crest to the medial malleolus [35] by means of an anthropometric tape measure (Seca 203, Hamburg, Germany). Once measured, the players were assessed following the protocol previously described by Linek et al. [35] and Iglesias-Caamaño et al. [4,31].

2.3.5. Single-Leg One-Repetition Maximum in Leg Press (1RM-SL)

The 1RM-SL test has been used to establish players' maximal concentric strength lateral symmetry. The players positioned themselves one-legged on a leg press ARTIS® (TECHNOGYM®, Cesena, CF, Italy) and pushed the platform until a full knee extension was reached. [31]. The test started with the hips aligned and at an initial 90° knee flexion.

2.3.6. Single-Leg Squat Jump (SL-SJ)

The SL-SJ test was used to establish the lateral symmetry of the players' explosive strength. A single-leg vertical jump was performed, as previously described by Iglesias-Caamaño et al. [31] in a MULTIPOWER® Smith machine (TECHNOGYM®, Cesena, CF, Italy). The test started from a 90° knee flexion [36] at 50% of the player's body weight. The average propulsive velocity of the bar displacement (m/s) was determined through a linear encoder (CHRONOJUMP Boscosystem®, Barcelona, Spain).

2.3.7. Triple Hop Test for Distance (THTD)

The THTD is a horizontal jumping test that consists of performing three consecutive single-leg jumps with each leg aiming to achieve the longest possible length [18]. Players' hands were fixed on the hips and the landing had to be balanced and firm, maintaining the position for 2–3 s. The jump length was measured with a large-distance tape-measure (Bellota 50022-30m, Bellota herramientas, s.l.u, Gipuzkoa, Spain) from the tip of the foot at the starting line to the heel of the third landing.

2.3.8. Radial Muscle Displacement (Dm)

Tensiomyography (TMG) was used to measure players' radial muscle belly displacement of the biceps femoris (BF) and rectus femoris (RF). TMG assessments were performed on both limbs following the protocol described by García-García et al. [37,38].

Radial muscle belly displacement was acquired using a digital displacement transducer (GK 30, Panoptik d.o.o., Ljubljana, Slovenia) set perpendicular to the thickest part of the muscle belly following Perotto's et al. [39] indications. The self-adhesive electrodes (5 × 5 cm, Cefar-Compex Medical AB Co., Ltd., Malmö, Sweden) were placed symmetrically at 5 cm from the sensor. An electrical stimulation was applied with a pulse duration of 1 ms and an initial current amplitude of 30 mA that was progressively increased in 10 mA steps until it reached 110 mA (maximal stimulator output). The electrical stimulus was produced by a TMG-S2 (EMF-FURLAN & Co. d.o.o., Ljubljana, Slovenia) stimulator. The percentage of lateral symmetry (LS) was calculated through algorithm (1), implemented by the TMG-BMC Tensiomyography® software.

$$Ls = 0.1 \cdot \frac{\min(TDr \cdot TDl)}{\max(TDr \cdot TDl)} + 0.6 \cdot \frac{\min(TCr \cdot TCl)}{\max(TCr \cdot TCl)} + 0.1 \cdot \frac{\min(TSr \cdot TSl)}{\max(TSr \cdot TSl)} + 0.2 \cdot \frac{\min(DMr \cdot DMl)}{\max(DMr \cdot DMl)} \quad (1)$$

2.4. Statistical Analyses

Adequacy sample size was calculated using G*Power v3.1.9.4 for Windows (Heinrich-Heine-Universität Düsseldorf, Germany) based on an effect size of 1.10, an alpha error of 0.05, and a power of 0.95. The univariate Kolmogorov–Smirnov test in conjunction with the Lilliefors test demonstrated the sample's normality. The homoscedasticity assumption was verified using Box's M test for equivalence of covariance, followed by a Tukey's HSD post hoc test. Relative reliability was calculated through intraclass correlation coefficient (ICC) analysis using single measurements, two-way mixed effects models, and absolute agreement. The coefficient of variation (CV) was used as a measure of absolute reproducibility. The influence on the magnitude and direction of asymmetry was examined by performing a two-way ANOVA using a test factor (within subjects) and a category factor (between subjects). Finally, an individual analysis of asymmetry was carried out establishing specific asymmetry thresholds following the formula %Asym + (0.2 × SD) [23] or the fixed asymmetry thresholds commonly used in scientific literature (10–15% difference between legs) [12,19]. Additionally, a two-way ANOVA was used to determine if the number of volleyball players classified as "asymmetrical" was modulated by the asymmetry threshold used and/or by the test. The effect sizes in the ANOVA two-way were reported as partial eta squared measurements ($\eta_p{}^2$) and interpreted as small (0.01), moderate (0.06), or large (0.14) [40]. An alpha level of $p < 0.05$ was considered statistically significant. All data were analyzed using SPSS v.24.0 for Windows (SPSS Inc., Chicago, IL, USA).

3. Results

Relative and absolute reliability values (ICC and CV respectively) were 0.92 and 2.2% for AKE, 0.89 and 3.1% for SL-CMJ, 0.77 and 2.5% for M-20Y, 0.93 and 2.6% for YBT, 0.87 and 13% for SL-SJ, 0.63 and 8.6% for THTD, and 0.98 and 3.8% for Dm.

The magnitude of asymmetry obtained by the volleyball players in each test based on their level of performance is shown in Table 2.

The analysis of variance results indicated large significant differences between tests when determining the magnitude of lateral asymmetry (F = 25.194; $p = 0.001$; $\eta_p{}^2 = 0.476$) but not between categories (F = 0.1951; $p = 0.164$; $\eta_p{}^2 = 0.008$) nor in test x category interaction (F = 0.961; $p = 0.473$; $\eta_p{}^2 = 0.033$).

The percentage of asymmetry calculated with the AKE was lower than with the SL-CMJ ($p = 0.001$), the 1RM-SL ($p = 0.001$), the SL-SJ ($p = 0.001$), the Dm of BF ($p = 0.001$), and the Dm of RF ($p = 0.001$). This percentage with SL-CMJ was of greater magnitude than the AKE and M-20Y ($p = 0.027$) and lower than the Dm of BF ($p = 0.001$) and RF ($p = 0.001$). The asymmetry detected with M-20Y was significantly lower than when using the SL-CMJ, the 1RM-SL ($p = 0.003$), the SL-SJ ($p = 0.004$), the Dm of BF ($p = 0.001$), and the Dm of RF ($p = 0.001$). Along this same line, the YBT obtained less asymmetry than the 1RM-SL ($p = 0.004$), SL-SJ ($p = 0.007$), Dm of BF ($p = 0.001$), and Dm of RF ($p = 0.001$).

The 1RM-SL obtained a greater asymmetry than the YTB ($p = 0.004$) and the THTD ($p = 0.024$); however, it was lower than when the Dm of BF ($p = 0.001$) and the Dm of RF ($p = 0.001$) were used. The percentage of asymmetry using the SL-SJ was also greater than with the YBT ($p = 0.007$) and the THTD ($p = 0.037$) but lower than with the Dm of BF ($p = 0.001$) and with the Dm of RF ($p = 0.001$). For the THTD, a lower percentage of lateral symmetry was obtained than the Dm of BF ($p = 0.001$) and Dm of RF ($p = 0.001$).

Finally, the Dm of BF and the Dm of RF obtained a magnitude of asymmetry greater than all other tests.

Table 2. Descriptive statistics of the magnitude of lateral asymmetry of the volleyball players.

Test	Category	Mean (%)	SD	IC 95%		Asymmetry Threshold
AKE	Junior	2.859	2.374	1.638	4.079	
	Senior	1.766	1.464	1.292	2.241	
	Total	2.098	1.838	1.606	2.590	2.46
SL-CMJ	Junior	9.636	4.567	1.709	20.981	
	Senior	8.451	7.565	5.626	11.276	
	Total	8.559	7.300	5.970	11.147	8.93
M-20Y	Junior	1.469	1.148	1.382	4.320	
	Senior	1.645	1.219	1.190	2.100	
	Total	1.629	1.197	1.205	2.053	1.87
YBT	Junior	4.904	3.230	3.243	6.565	
	Senior	3.180	3.494	2.031	4.328	
	Total	3.713	3.479	2.772	4.653	4.41
1RM-SL	Junior	9.627	8.775	2.291	16.963	
	Senior	13.319	3.847	8.542	18.097	
	Total	11.047	7.304	6.634	15.461	12.51
SL-SJ	Junior	12.387	10.280	3.793	20.981	
	Senior	9.522	12.515	6.018	25.061	
	Total	11.285	10.768	4.778	17.792	22.25
THTD	Junior	4.056	3.760	1.367	6.746	
	Senior	5.621	8.474	1.464	12.705	
	Total	4.751	6.139	1.698	7.804	5.98
Dm.BF	Junior	29.804	24.110	13.607	46.001	
	Senior	20.997	11.919	9.974	32.021	
	Total	26.379	20.288	16.290	36.468	30.44
Dm.RF	Junior	30.275	15.330	19.976	40.574	
	Senior	22.582	11.710	11.752	33.412	
	Total	27.283	14.196	20.224	34.343	30.12

AKE: active knee extension test, SL-CMJ: single-leg countermovement test, M-20Y: modified 20-yard shuttle run test, YBT: Y-balance test, 1RM-SL: single-leg one-repetition maximum test in leg press, SL-SJ: single-leg squat jump test, THTD: triple hop test for distance, Dm.BF: radial muscle belly displacement of the biceps femoris, Dm.RF: radial muscle belly displacement of the rectus femoris, SD: standard deviation.

The ANOVA results were similar to those shown previously in the analysis of the magnitude (without the direction), finding moderately significant differences between tests when determining the magnitude and direction of the lateral asymmetry (F = 1.961; $p = 0.044$; $\eta_p^2 = 0.066$), but not between categories (F = 0.297; $p = 0.586$; $\eta_p^2 = 0.001$). However, when we analyzed the magnitude and the direction jointly, the interaction test × category (F = 2.499; $p = 0.009$; $\eta_p^2 = 0.083$) indicated a moderately significant difference.

As can be seen in Figure 2, in the AKE, the SL-CMJ, and the THTD, the asymmetry was towards the NDL in both categories (less performance in the right leg). The 1RM-SL, the SL-SJ, and the Dm of BF showed an asymmetry towards the DL (less performance in the left leg). The M-20Y, the YBT, and the Dm of RF showed different orientations depending on the evaluated category (junior vs. senior).

For the category × test interaction, a greater lateral asymmetry was found in the seniors' 1RM-SL (−13.32% vs. −2.75%) and SL-SJ (−8.05% vs. −1.74%) and a greater asymmetry in junior players was found in the Dm of BF (−15.86% vs. −3.99%) and the Dm of RF (−18.24% vs. 6.89%).

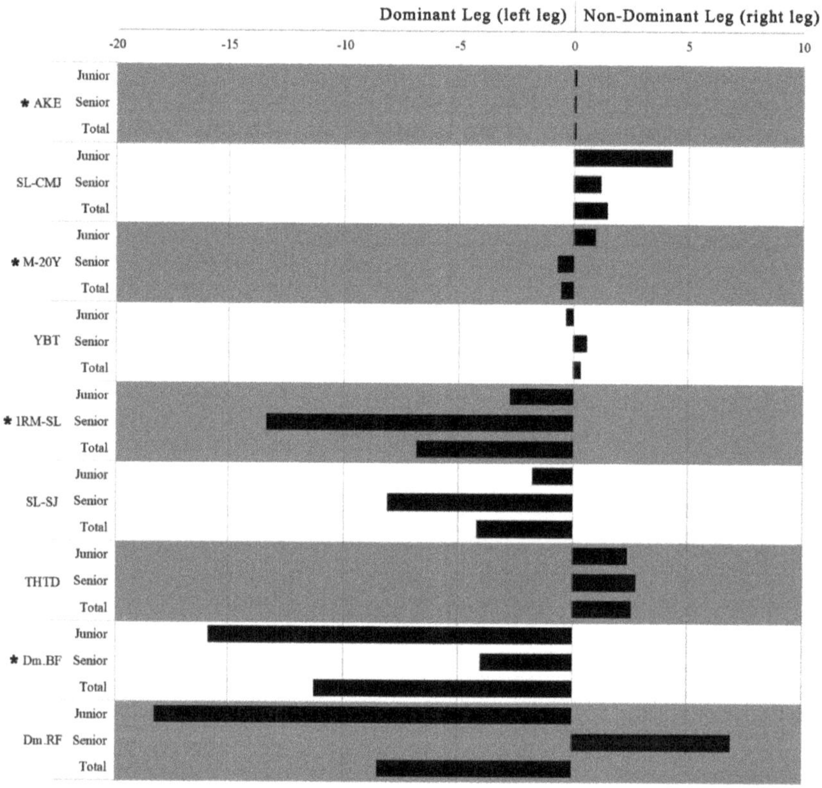

Figure 2. Magnitude of lateral asymmetry of the volleyball players obtained in each test. A positive percentage indicates a direction of asymmetry towards the DL (left leg) and a negative percentage indicates an asymmetry towards the NDL (right leg). *: Significant ($p < 0.05$); AKE: active knee extension test; SL-CMJ: single-leg countermovement test; M-20Y: modified 20-yard shuttle run test; YBT: Y-balance test; 1RM-SL: single-leg one-repetition maximum test in leg press; SL-SJ: single-leg squat jump test; THTD: triple hop test for distance; Dm.BF: radial muscle belly displacement of the biceps femoris; Dm.RF: radial muscle belly displacement of the rectus femoris; DL: dominant leg, NDL: non-dominant leg.

The individualized analysis for each player based on each test asymmetry threshold, using (a) the specific asymmetry threshold formula or (b) the 10% fixed asymmetry thresholds, shows that there are not any significant differences in the determination of volleyball players' lateral asymmetry (F = 0.015; $p = 0.903$; $\eta_p^2 = 0.001$). However, there are significant differences in the determination of volleyball players' lateral asymmetry in the threshold × test interaction (F = 4.404; $p = 0.004$; $\eta_p^2 = 0.662$) with a large effect size. In addition, there are differences in test (F = 3.266; $p = 0.018$; $\eta_p^2 = 0.592$) that also show a large effect size.

Specifically, when using the specific asymmetry threshold formula, the percentage of classified players as asymmetrical is between 15.38% for the SL-SJ test and 42.86% for the YBT test. However, if the fixed threshold of 10% difference between limbs is used, the percentage of asymmetrical players significantly varies between 0% (AKE and M-20Y tests) and 88.89% (Dm of RF test).

4. Discussion

Our main findings indicate that the type of test used determines the magnitude and direction of asymmetry, and the asymmetry thresholds of well-trained volleyball players. Furthermore, when analyzing a player's category together with the test used, there is an interaction that also influences the magnitude and direction of asymmetry. The 1RM-SL, SL-SJ, and Dm of BF showed a clear asymmetry towards the DL, that is, the left leg had an less performance. In contrast, the AKE, SL-CMJ, and THTD showed asymmetry towards the NDL, being the right leg, which had less performance. Very variable magnitudes of asymmetry have been found between the tests that composed our battery (1.46–30.27%). The individualized analysis using specific asymmetry thresholds again reveals that the type of test performed modulates the percentage of players considered asymmetrical, ranging between 15.38% in the SL-SJ and 42.82% in the YBT. When using the 10% fixed threshold, the values hover between 0% for the AKE and M-20Y tests and 88.89% for the Dm of RF.

The ICC and CV values obtained were good to excellent for all tests, except for THTD's relative reliability, which was moderate. This corroborates the good reliability of this battery test that Iglesias-Caamaño et al. [4,31] had previously reported.

Overall, these findings are in line with those recently reported by Kozinc and Šarabon [41] with young volleyball players. These authors also found a highly variable magnitude of asymmetry (2.0–31.2%) among their wide battery tests. In addition, their magnitudes of asymmetry were similar to ours in the tests that they had in common with our testing battery (COD: 1.62 ± 1.19% vs. 2.02 ± 1.83%; SL-CMJ: 8.56 ± 7.3% vs. 10.24 ± 9.39%; THTD: 4.7 5 ± 6.14% vs. 4.3 ± 2.8%). Therefore, with due prudence these data could serve as a reference of asymmetry for volleyball players.

Regarding the magnitude of asymmetry found in soccer players, it seems that the asymmetry shown in well-trained volleyball players does not differ significantly through an isokinetic test in concentric and eccentric contractions [42] nor those evaluated by a unilateral 1RM (9.9 ± 7.2% vs. 11.04 ± 7.3%) [43]. Similarly, our volleyball players did not differ from the asymmetry values revealed by recreational rugby and soccer players assessed through a SL-CMJ (7.2 ± 6.1% vs. 8.56 ± 7.3%) [25], nor those reported by Meylan et al. [19] showing an asymmetry index of 6.0% in 30 team athletes. Along these lines, our findings indicate similar values of asymmetry to netball players in COD tests [44], although in a different test (505 agility test: 2.3 ± 2.3% vs. M-20Y 1.69 ± 1.2%). In addition, our findings in the YBT are in line with those of Ryu et al. [45] in professional baseball players, finding no significant differences in the analysis of direction between DL and NDL. However, volleyball players appear to have less asymmetry in active knee mobility (AKE) than Gaelic football players [7], where they observed 5.5 ± 4.8% asymmetry in their 570 players compared to the 2.1 ± 1.84% asymmetry observed in our 29 volleyball players. These differences between volleyball and Gaelic football players may be due to their differences in the mechanical performance model, since volleyball players perform actions in much smaller spaces with less pronounced changes of direction and in lower defensive positions than Gaelic football players. Regarding the Dm of the BF and RF, it seems that our volleyball players show higher values of inter-limb asymmetry (26.38 ± 20.29% and 27.28 ± 14.20%, respectively) than those reported in elite futsal players [46]. In contrast, Rodríguez-Ruiz et al. [47] found no significant inter-limb differences in these parameters in professional volleyball players, nor did García-García et al. [48] in professional soccer players. However, the latter authors pointed out that these asymmetries can vary throughout the soccer player's season. Notwithstanding, asymmetry in these studies was not calculated using an index, so these findings should be compared with caution. Therefore, these differences between authors could be to a certain extent due to the methodological differences in data collection to reach the maximum Dm, the different time of the season, and the differences in the mechanical performance model (i.e., ball kick in soccer vs. futsal).

The determination of the magnitude together with the direction of asymmetry requires a more in-depth analysis to be able to correctly interpret the findings. First, it is necessary to establish an appropriate criterion to determine which is the DL of the athlete, as it will mark the direction of magnitude towards their DL or NDL in the evaluated task [5]. This is because it seems that the dominance in limbs is rarely the same between different tasks, as it has been suggested in recreational athletes [5] or between tests such as THTD and CODs [16]. In these lines, our findings indicate that the 1RM-SL, SL-SJ, and Dm of the BF show a clear asymmetry towards the DL, that is, the performance of the DL should be improved to reduce asymmetry. This lateral asymmetry is possibly due to the need to produce great levels of maximum strength in the 1RM-SL and SL-SJ and the high muscle tone in the Dm of BF of the NDL. Nevertheless, the AKE, SL-CMJ, and THTD clearly show an asymmetry towards the NDL, that is, the performance of the NDL should be improved to reduce asymmetry. This lateral asymmetry is probably caused by the need to exert explosive and reactive force during a greater ROM with the DL in the volleyball mechanical model. In view of these circumstances, it seems advisable to determine the dominant limb of each player for each test performed.

In relation to the magnitude and direction of asymmetry based on category, small differences have been observed in terms of direction between categories. Tasks such as the M-20Y showed asymmetry towards the NDL while the YBT and Dm of RF towards the DL in seniors. Hence, it is possible that specific volleyball adaptions may create asymmetries in lower limbs over the years. Therefore, the years of practice seem to influence an increase in asymmetry in jumping and strength tasks (i.e., 1RM and SL-SJ); however, at younger ages these differences seem to appear in muscle tone.

The limitations that arise in the interpretation of the magnitude and the magnitude and direction of asymmetry analyzed together suggest that an individual approach is necessary for each player to determine whether there is a lateral asymmetry. Traditionally, in the scientific literature it has been suggested that an inter-limb difference of >10–15% implies an asymmetry [49]. However, it has recently been pointed out that the threshold to consider an athlete asymmetrical should be established based on the sample assessed and the type of test carried out [5]. In fact, it has been suggested that the commonly accepted >10% threshold for classifying individuals as asymmetrical should be reconsidered and reinvestigated [41]. This individualization would allow those interested to really determine in which motor skills the player is asymmetrical. Our findings indicate that different results are obtained depending on the asymmetry criterion, whether using the fixed asymmetry threshold of >10–15% inter-limb difference or the specific asymmetry threshold for each category of athletes and each test carried out. For example, the AKE test did not show asymmetries >10% for any player, but if we use the specific asymmetry threshold considering the category and the test, 31.03% of the players would be classified as asymmetrical. Similarly, Dos'Santos et al. [44] detected a 30% fraction of asymmetrical athletes with a specific asymmetry threshold for the 505 agility test, while using the fixed asymmetry threshold >10% only a 5% of athletes these were identified. In fact, Kozinc and Šarabon [41] found no asymmetries in volleyball players when applying the fixed asymmetry threshold (>10%) in two agility tests with 180° and 90° turns, as in our study with the M-20Y. However, the number of asymmetrical players found in our study was 41.67% if we used that specific asymmetry threshold.

Several limitations must be considered when interpreting our findings. On the one hand, the low number of volleyball players in our sample does not allow for reference data to be collected. Therefore, a representative sample of volleyball players is needed to obtain reference thresholds. On the other hand, the lack of female players in our sample limits our findings only to male volleyball players, so it is undoubtedly necessary to address gender differences in the future.

5. Conclusions

In conclusion, the type of test used determines the magnitude and direction of asymmetry and the asymmetry thresholds of well-trained volleyball players. It seems necessary to carry out a variety of specific volleyball tests that assess strength, power, agility, ROM, balance, and especially, sport-specific technical gestures (such as the attack hit in volleyball) to globally understand a player's asymmetry. In addition, it also seems necessary to establish an individual asymmetry analysis for each player based on a specific asymmetry threshold for the athlete's category and the test used. In this sense, reference values are required for specific asymmetry thresholds to be able to classify the volleyball players. This would help establish the necessary training strategies (strength, flexibility, etc.) to improve the imbalance. Furthermore, this could be transferred to other sports through an appropriate specific testing battery that includes tests that assess certain specific aspects related to the performance factors of the sport in question.

Author Contributions: Conceptualization, M.I.-C. and Ó.G.-G.; methodology, M.I.-C., T.Á.-Y. and A.C.-D.; software, J.C.-L.; formal analysis, Ó.G.-G. and M.I.-C.; investigation, Ó.G.-G., M.I.-C., T.Á.-Y., A.C.-D. and J.C.-L.; resources, T.Á.-Y. and J.C.-L.; writing—original draft preparation, M.I.-C., T.Á.-Y. and Ó.G.-G.; writing—review and editing, Ó.G.-G., M.I.-C., T.Á.-Y. and A.C.-D.; supervision, Ó.G.-G. All authors have read and agreed to the published version of the manuscript.

Funding: This research received no external funding.

Institutional Review Board Statement: The study was conducted in accordance with the Declaration of Helsinki, and approved by the Ethics Research Committee of the Faculty of Education and Sports Sciences of the University of Vigo (Protocol code 06-1019 approved October 2019).

Informed Consent Statement: Informed consent was obtained from all subjects involved in the study.

Data Availability Statement: The data that support the findings of this study are available from the authors M.I.-C. and O.G.-G., upon request.

Conflicts of Interest: The authors declare no conflict of interest.

References

1. Viitasalo, J.T.; Rusko, H.; Pajala, O.; Rahkila, P.; Ahila, M.; Montonen, H. Endurance requirements in volleyball. *Can. J. Sport Sci.-Rev. Can. Des. Sci. Du Sport* **1987**, *12*, 194–201.
2. Fontani, G.; Ciccarone, G.; Giulianini, R. Nuove regole di gioco ed impegno fisico nella pallavolo. *SDS* **2000**, *19*, 14–20.
3. Sheppard, J.; Gabbett, T.; Riggs, M. Volleyball. In *Physiological Test for Elite Athletes*, 2nd ed.; Human Kinetics: Champaign, IL, USA, 2013; pp. 475–486.
4. Iglesias-Caamano, M.; Alvarez-Yates, T.; Carballo-Lopez, J.; Cuba-Dorado, A.; Garcia-Garcia, O. Interday Reliability of a Testing Battery to Assess Lateral Symmetry and Performance in Well-Trained Volleyball Players. *J. Strength Cond. Res.* **2022**, *36*, 895–901. [CrossRef] [PubMed]
5. Bishop, C.; Lake, J.; Loturco, I.; Papadopoulos, K.; Turner, A.; Read, P. Interlimb Asymmetries: The Need for an Individual Approach to Data Analysis. *J. Strength Cond. Res.* **2021**, *35*, 695–701. [CrossRef]
6. Hewit, J.K.; Cronin, J.B.; Hume, P.A. Asymmetry in multi-directional jumping tasks. *Phys. Ther. Sport* **2012**, *13*, 238–242. [CrossRef]
7. Newton, R.U.; Gerber, A.; Nimphius, S.; Shim, J.K.; Doan, B.K.; Robertson, M.; Pearson, D.R.; Craig, B.W.; Hakkinen, K.; Kraemer, W.J. Determination of functional strength imbalance of the lower extremities. *J. Strength Cond. Res.* **2006**, *20*, 971–977. [PubMed]
8. Hadzic, V.; Sattler, T.; Markovic, G.; Veselko, M.; Dervisevic, E. The isokinetic strength profile of quadriceps and hamstrings in elite volleyball players. *Isokinet. Exerc. Sci.* **2010**, *18*, 31–37. [CrossRef]
9. Wagner, H.; Tilp, M.; von Duvillard, S.P.; Mueller, E. Kinematic Analysis of Volleyball Spike Jump. *Int. J. Sports Med.* **2009**, *30*, 760–765. [CrossRef] [PubMed]
10. Lobietti, R.; Coleman, S.; Pizzichillo, E.; Merni, F. Landing techniques in volleyball. *J. Sports Sci.* **2010**, *28*, 1469–1476. [CrossRef] [PubMed]
11. Hernandez-Hernandez, E.; Montalvo-Espinosa, A.; Garcia-de-Alcaraz, A. A Time-Motion Analysis of the Cross-Over Step Block Technique in Volleyball: Non-Linear and Asymmetric Performances. *Symmetry* **2020**, *12*, 1027. [CrossRef]
12. Impellizzeri, F.M.; Rampinini, E.; Maffiuletti, N.; Marcora, S.M. A vertical jump force test for assessing bilateral strength asymmetry in athletes. *Med. Sci. Sports Exerc.* **2007**, *39*, 2044–2050. [CrossRef] [PubMed]
13. Atkins, S.J.; Bentley, I.; Hurst, H.T.; Sinclair, J.K.; Hesketh, C. The Presence of Bilateral Imbalance of the Lower Limbs in Elite Youth Soccer Players of Different Ages. *J. Strength Cond. Res.* **2016**, *30*, 1007–1013. [CrossRef] [PubMed]
14. Raya-González, J.; Beato, M.; Castillo, D. Differences in power performance attending to sprint capacity and asymmetry level in young elite soccer players. In *An Essential Guide to Sports Performance*; Nova Science Publishers: New York, NY, USA, 2020; pp. 79–98. ISBN 9781536176087.
15. Bailey, C.; Sato, K.; Alexander, R.; Chiang, C.-Y.; Stone, M.H. Isometric force production symmetry and jumping performance in collegiate athletes. *J. Trainology* **2013**, *2*, 1–5. [CrossRef]

16. Dos'Santos, T.; Thomas, C.; Jones, P.A.; Comfort, P. Asymmetries in single and triple hop are not detrimental to change of direction speed. *J. Trainology* **2017**, *6*, 35–41. [CrossRef]
17. Loturco, I.; Pereira, L.A.; Kobal, R.; Abad, C.C.C.; Komatsu, W.; Cunha, R.; Arliani, G.; Ejnisman, B.; Pochini, A.D.; Nakamura, F.Y.; et al. Functional Screening Tests: Interrelationships and Ability to Predict Vertical Jump Performance. *Int. J. Sports Med.* **2018**, *39*, 189–197. [CrossRef]
18. Hamilton, R.T.; Shultz, S.J.; Schmitz, R.J.; Perrin, D.H. Triple-hop distance as a valid predictor of lower limb strength and power. *J. Athl. Train.* **2008**, *43*, 144–151. [CrossRef]
19. Meylan, C.M.P.; Nosaka, K.; Green, J.; Cronin, J.B. Temporal and kinetic analysis of unilateral jumping in the vertical, horizontal, and lateral directions. *J. Sports Sci.* **2010**, *28*, 545–554. [CrossRef]
20. Parkinson, A.O.; Apps, C.L.; Morris, J.G.; Barnett, C.T.; Lewis, M.G.C. The Calculation, Thresholds and Reporting of Inter-Limb Strength Asymmetry: A Systematic Review. *J. Sports Sci. Med.* **2021**, *20*, 594–617. [CrossRef] [PubMed]
21. Brumitt, J.; Heiderscheit, B.C.; Manske, R.C.; Niemuth, P.E.; Rauh, M.J. Lower extremity functional tests and risk of injury in division iii collegiate athletes. *Int. J. Sports Phys. Ther.* **2013**, *8*, 216–227.
22. Bishop, C. Interlimb Asymmetries: Are Thresholds a Usable Concept? *Strength Cond. J.* **2021**, *43*, 32–36. [CrossRef]
23. Dos'Santos, T.; Thomas, C.; Jones, P.A. Assessing Interlimb Asymmetries: Are We Heading in the Right Direction? *Strength Cond. J.* **2021**, *43*, 91–100. [CrossRef]
24. Hopkins, W.G. How to interpret changes in an athletic performance test. *Sportscience* **2004**, *8*, 1–7.
25. Bishop, C.; Read, P.; Chavda, S.; Jarvis, P.; Turner, A. Using Unilateral Strength, Power and Reactive Strength Tests to Detect the Magnitude and Direction of Asymmetry: A Test-Retest Design. *Sports* **2019**, *7*, 58. [CrossRef] [PubMed]
26. Exell, T.A.; Irwin, G.; Gittoes, M.J.R.; Kerwin, D.G. Implications of intra-limb variability on asymmetry analyses. *J. Sports Sci.* **2012**, *30*, 403–409. [CrossRef]
27. Mattes, K.; Wollesen, B.; Manzer, S. Asymmetries of maximum trunk, hand, and leg strength in comparison to volleyball and fitness athletes. *J. Strength Cond. Res.* **2018**, *32*, 57–65. [CrossRef]
28. Schons, P.; Da Rosa, R.G.; Fischer, G.; Berriel, G.P.; Fritsch, C.G.; Nakamura, F.Y.; Baroni, B.M.; Peyre-Tartaruga, L.A. The relationship between strength asymmetries and jumping performance in professional volleyball players. *Sports Biomech.* **2019**, *18*, 515–526. [CrossRef]
29. Fort-Vanmeerhaeghe, A.; Gual, G.; Romero-Rodriguez, D.; Unnitha, V. Lower Limb Neuromuscular Asymmetry in Volleyball and Basketball Players. *J. Hum. Kinet.* **2016**, *50*, 135–143. [CrossRef]
30. Stephens, T.M.; Lawson, B.R.; Reiser, R.F. Bilateral asymmetries in max effort single-leg vertical jumps. *Biomed. Sci. Instrum.* **2005**, *41*, 317–322.
31. Iglesias-Caamano, M.; Carballo-Lopez, J.; Alvarez-Yates, T.; Cuba-Dorado, A.; Garcia-Garcia, O. Intrasession Reliability of the Tests to Determine Lateral Asymmetry and Performance in Volleyball Players. *Symmetry* **2018**, *10*, 416. [CrossRef]
32. Gajdosik, R.; Lusin, G. Hamstring muscle tightness: Reliability of an active-knee-extension test. *Phys. Ther.* **1983**, *63*, 1085–1088. [CrossRef]
33. Thomas, C.; Comfort, P.; Dos'Santos, T.; Jones, P.A. Determining Bilateral Strength Imbalances in Youth Basketball Athletes. *Int. J. Sports Med.* **2017**, *38*, 683–690. [CrossRef] [PubMed]
34. Sekulic, D.; Spasic, M.; Mirkov, D.; Cavar, M.; Sattler, T. Gender-specific influences of balance, speed, and power on agility performance. *J. Strength Cond. Res.* **2013**, *27*, 802–811. [CrossRef] [PubMed]
35. Linek, P.; Sikora, D.; Wolny, T.; Saulicz, E. Reliability and number of trials of Y Balance Test in adolescent athletes. *Musculoskelet. Sci. Pract.* **2017**, *31*, 72–75. [CrossRef]
36. Kockum, B.; Heijne, A. Hop performance and leg muscle power in athletes: Reliability of a test battery. *Phys. Ther. Sport* **2015**, *16*, 222–227. [CrossRef] [PubMed]
37. Garcia-Garcia, O.; Cancela-Carral, J.M.; Martinez-Trigo, R.; Serrano-Gomez, V. Differences in the contractile properties of the knee extensor and flexor muscles in professional road cyclists during the season. *J. Strength Cond. Res.* **2013**, *27*, 2760–2767. [CrossRef] [PubMed]
38. Garcia-Garcia, O.; Serrano-Gomez, V.; Hernandez-Mendo, A.; Tapia-Flores, A. Assessment of the in-season changes in mechanical and neuromuscular characteristics in professional soccer players. *J. Sports Med. Phys. Fit.* **2016**, *56*, 714–723.
39. Perotto, A.; Delagi, E.F.; Iazzetti, J.; Morrison, D. *Anatomical Guide for the Electromyographer: The Limbs and Trunk*; Charles, C., Ed.; Thomas Ltd.: Springfield, IL, USA, 2005; ISBN 978-0-398-08649-7.
40. Cohen, J. *Statistical Power Analysis for the Behavioral Sciences*, 2nd ed.; Lawrence Erlbaum: New York, NY, USA, 1988. [CrossRef]
41. Kozinc, Z.; Sarabon, N. Inter-Limb Asymmetries in Volleyball Players: Differences between Testing Approaches and Association with Performance. *J. Sports Sci. Med.* **2020**, *19*, 745–752.
42. Magalhaes, J.; Oliveira, J.; Ascensao, A.; Soares, J. Concentric quadriceps and hamstrings isokinetic strength in volleyball and soccer players. *J. Sports Med. Phys. Fit.* **2004**, *44*, 119–125.
43. Sannicandro, I.; Piccinno, A.; Rosa, R.; De Pascalis, S. Functional asymmetry in the lower limb professional soccer players. *Br. J. Sports Med.* **2011**, *45*, 370. [CrossRef]
44. Dos'Santos, T.; Thomas, C.; Jones, P.A.; Comfort, P. Assessing asymmetries in change of direction speed performance: Application of change of direction deficit. *J. Strength Cond. Res.* **2019**, *33*, 2953–2961. [CrossRef]

45. Ryu, C.H.; Park, J.; Kang, M.; Oh, J.H.; Kim, Y.K.; Kim, Y.I.; Lee, H.S.; Seo, S.G. Differences in lower quarter Y-balance test with player position and ankle injuries in professional baseball players. *J. Orthop. Surg.* **2019**, *27*. [CrossRef] [PubMed]
46. Lopez-Fernandez, J.; Garcia-Unanue, J.; Sanchez-Sanchez, J.; Colino, E.; Hernando, E.; Gallardo, L. Bilateral Asymmetries Assessment in Elite and Sub-Elite Male Futsal Players. *Int. J. Environ. Res. Public Health* **2020**, *17*, 3169. [CrossRef] [PubMed]
47. Rodriguez-Ruiz, D.; Rodriguez-Matoso, D.; Quiroga, M.E.; Sarmiento, S.; Garcia-Manso, J.M.; Da Silva-Grigoletto, M.E. Study of mechanical characteristics of the knee extensor and flexor musculature of volleyball players. *Eur. J. Sport Sci.* **2012**, *12*, 399–407. [CrossRef]
48. García-García, O.; Serrano-Gómez, V.; Cuba-Dorado, A. Evolution of the lateral symmetry of the lower limbs of professional footballers during the season. In Proceedings of the IV NSCA International Conference, Murcia, Spain, 26–28 June 2014.
49. Barber, S.D.; Noyes, F.R.; Mangine, R.E.; McCloskey, J.W.; Hartman, W. Quantitative assessment of functional limitations in normal and anterior cruciate ligament-deficient knees. *Clin. Orthop. Relat. Res.* **1990**, *255*, 204–214. [CrossRef]

symmetry

MDPI

Article

Born to Score? The Relationship between Left-Handedness and Success from the 7-Meter Line

Aron Laxdal [1,*], Andreas Ivarsson [1,2], Sveinn Thorgeirsson [3,4] and Tommy Haugen [1]

1 Department of Sport Science and Physical Education, University of Agder, 4630 Kristiansand, Norway
2 Center of Research on Welfare, Health and Sport, Halmstad University, 30118 Halmstad, Sweden
3 Physical Activity, Physical Education, Sport and Health Research Centre, Reykjavik University, 102 Reykjavik, Iceland
4 Faculty of Kinesiology, University of Split, 21000 Split, Croatia
* Correspondence: aron.laxdal@uia.no; Tel.: +47-45265312

Abstract: An asymmetry in the prevalence of left-handedness vs. right-handedness in society has supposedly resulted in negative frequency-dependent advantages for left-handers in interactive sports. The aim of this study was to test whether these advantages apply to handball by examining whether being left-handed is beneficial when executing 7 m shots, a highly unimanual movement. All 1,625 7 m shots at the men's 2016–2022 European championships were analyzed using a Bayesian two-level analysis. While the results did not indicate that left-handers were more likely to score from any single 7 m shot, left-handers were overrepresented among the designated shooters compared to both the population as a whole (38% vs. 11.6%) and left-handers on any given handball team (38% vs. 25%). The implication here was that handedness plays no role in the outcome of 7 m shots at the world-class level, but handedness does appear to play a role in who becomes a world-class 7 m shooter.

Keywords: asymmetry; handball; laterality; negative frequency-dependent advantage; penalties; throwing arm

1. Introduction

Approximately 10% of the world is left-handed, with some variations depending on sex and context [1]. While being left-handed has been linked with negative health outcomes [1,2], it has also been found to be surprisingly advantageous in sports [3,4]. The innate characteristic appears to be especially advantageous in interactive sports where an athlete must react to an incoming object, such as boxing, mixed martial arts, fencing, tennis, cricket, and volleyball [3–5]. However, the degree of advantage appears to vary greatly depending on various factors, resulting in a varying prevalence of left-handedness (as high as 37–48% in judo and fencing, and as low as 13% in badminton and baseball [4]).

The mechanism believed to explain this tendency is generally referred to as the negative frequency-dependent advantage [4]. It concerns the reduced ability to correctly anticipate the actions of those who deviate from the expected norm. Most of the world population is right-handed, so most of our learned perceptual expertise and a database of knowledge is based on interactions with, and observations of, right-handers [4,6]. The lower perceptual familiarity with the movement patterns of the relatively rarer left-handers can therefore affect the visual information processes of the would-be opponents [7]. For some reason, this advantage has been found to be greater when it comes to hands than feet [8].

Handball makes for a particularly interesting context to study the negative frequency-dependent advantage for two reasons. Firstly, the sport is highly unimanual and using the less preferred arm is practically unheard of, in contrast to sports such as boxing, mixed martial arts, and soccer that are more bimanual and -pedal. Secondly, left-handers have a distinct tactical advantage over right-handers on the right side of the court due to the available shooting angles and the movement patterns required to create high-quality goalscoring opportunities [9,10]. Being left-handed will therefore offer players an advantage in terms of selection, resulting in an overrepresentation compared to the general public [9,10]. In fact, most teams aim to have two left-handers on the field at any given time and four left-handers on the squad.

While handball is a relatively complex interactive sport, which makes studying negative frequency-dependent effects during regular gameplay difficult, the 7 m shot represents a relatively controlled situation where two players duel against each other without direct interference. The 7 m shot itself is an uncontested throw from the 7 m line, which is given when a player is robbed of a clear goalscoring opportunity in an unauthorized manner. In contrast to sports such as basketball and ice hockey, the fouled player is not required to take the shot themselves. The shot type typically has a higher success rate than regular gameplay [11], meaning that shooter selection should be carried out with deliberation.

As the 7 m line is placed centrally on the field, any angle-related advantages that the players might otherwise have, irrespective of handedness, are wiped out. The best 7 m shooters on any given team should therefore, at least theoretically, be selected purely based on performance and not for any tactical or strategic reasons. It is therefore quite interesting that Lobinger et al. [12] found that 43% of successful 7 m shots at the 2010 men's European Championship were taken by left-handers. While no clear explanations were offered, the ratio is in no proportion to the left-handers in the general population or those on the handball field at any given time.

When attempting to save a 7 m shot, the goalkeeper is free to move between the 4 m line and the goal line, with most goalkeepers opting to take a high position to reduce the angle of the shot. Due to the proximity between the two opposing parties, the goalkeeper will have to anticipate the shooter's movements [13,14]. Reacting to the trajectory of the ball in real time would be almost impossible, so the goalkeeper has to rely on various subtle cues and kinematic knowledge to predict where the shooter is likely to place the ball [14,15].

Previous research on goalkeepers and their efforts to anticipate the trajectory of 7 m throws has indicated that goalkeepers perform worse anticipating the throws of left-handers compared to right-handers [7]. Interestingly, the goalkeepers' gaze behavior and response time were the same irrespective of handedness, indicating that the discrepancy is due to problems categorizing or interpreting the information correctly [7,16]. This appears to indicate that some kind of negative perceptual frequency effect is at play during 7 m throws.

The aim of this study was therefore to examine (1) the prevalence of left-handed 7 m shooters and (2) the relationship between handedness and scoring from 7 m shots. The expectation was that left-handers would be overrepresented compared to both the general population and the population of handball players, and that they would be more likely to score from the 7 m line than their right-handed counterparts.

2. Methods

2.1. Procedure and Data

All 7 m shots at the four European championships for men between 2016 and 2022 were observed and logged into the official play-by-play match reports by representatives from the European Handball Federation. A total of 185 different players from 27 teams executed a total of 1625 7-m shots in 229 games ($M_{shots} = 7.1$, SD = 2.86).

The identity of the shooter, the outcome of the shot, and the score at the time of the 7 m shot were retrieved from the play-by-play data. Player height was gathered using the official webpage of the European handball federation and handedness was assessed using information from the same webpage in addition to image and video searches. The total number of 7 m shots during each game and the ratio between goals and misses were cross-referenced with the official match reports from each game to validate the information.

As the data that were used in this study are openly available online and are part of the public domain, no ethical approval was needed, and informed consent was not obtained from the players.

2.2. Statistical Analyses

Binomial tests of proportions were performed to determine whether left-handers were overrepresented within the sample. In line with previous studies on negative frequency-dependent advantages in interactive sports [17,18], the study sample was tested against a conservative estimate of expected left-handed males in the population (11.6% [1]). Additionally, the number of left-handed 7 m shooters was compared to the actual prevalence of left-handers at the four championships (25%; see Supplementary File S1).

The relationship between handedness and scoring from 7 m shots was analyzed using a Bayesian two-level analysis in Mplus 8.4. To account for dependency within the data, the outcomes of the 7 m shots (level one) were nested within the shooters (level two), with handedness being included as a covariate on level two. The height of the shooter and the goal difference at the time of the 7 m shot were controlled for on level one. If the 7 m shot had missing data on independent variables, it was excluded from this analysis, giving a total of 158 players and 1,615 shots. A Monte Carlo simulation, following the guidelines from Muthén and Muthén [19] was performed to determine sufficient power for the study. Specifying small effect sizes for all estimated relationships between independent and dependent variables, the results showed that 1,615 shots spread across 158 individual shooters yielded sufficient power (Beta > 0.80) to detect an effect.

A Bayesian estimator was used for the two-level analysis (for a comparison between the Bayesian estimator and the more traditional frequentist estimator, see Wagenmakers et al. [20]) and Markov Chain Monte Carlo simulation procedures with a Gibbs sampler with 200,000 iterations were performed. A potential scale-reduction factor of around 1 was used to indicate adequate convergence. Model fit was assessed using the posterior predictive p (PPp) value and its accompanying 95% confidence interval [21]. Credibility intervals (CI) were estimated for all parameters within the models. The null hypothesis was rejected if the 95% CI did not include zero [22].

3. Results

Of the 1,625 7 m shots, 1,224 (75.3%) resulted in a goal and 2,401 (24.7%) resulted in a miss. A total of 890 throws (54.8%) were performed with the right hand, while 735 throws (45.2%) were performed with the left hand. Out of the 158 different shooters, 98 were right-handed (62%), and 60 were left-handed (38%). The results of the binomial tests of proportions indicated that left-handers were overrepresented compared to both the general population (38% vs. 11.6%; $p < 0.001$) and handball teams in general (38% vs. 25%; $p < 0.001$).

The results from the Bayesian two-level analysis showed a good model fit to the data (PPp = 0.53, 95% CI = −10.99, 10.74). The included variables explained 0.5% of the variance of scoring on level one and 3.4% on level two, respectively. Handedness had no credible relationship with scoring ($\beta = 0.14$, 95% Confidence Interval = −0.32, 0.59). The other two covariates had no relationship with the likelihood of scoring.

4. Discussion

This study examined whether left-handed shooters were overrepresented as 7 m shooters in men's elite handball. While an overrepresentation of left-handers in handball compared to the general population is to be expected, left-handed 7 m shooters were also overrepresented compared to the prevalence of left-handers at the championships in question. These results are in line with findings from other contexts, where left-handed boxers, MMA fighters, fencers, and cricket bowlers have been found to be overrepresented at both the elite and amateur level [3,5,23,24]. Congruent with the findings of Lobinger et al. [12], almost half of the 7 m shots at the four championships were executed by left-handed shooters. This was disproportionate to the ratio of right- vs. left-handed 7 m shooters at the four championships, indicating that left-handers take a disproportionate number of 7 m shots compared to their right-handed counterparts.

While limited weight can be placed on anecdotal evidence, the last four tiebreaking 7 m shootouts at the highest level of handball (Champions League semi-final in 2014, Champions League final in 2016, World Championship quarter-final in 2021 and Champions League final in 2022) have also seen an overrepresentation of left-handed shooters. A total of 16 out of the 38 (42%) 7 m shooters were left-handed (see Supplemental File S2), adding further support to left-handers being overrepresented at the 7 m line at the elite level.

This study also aimed to determine whether left-handed players were more likely to score from the 7 m line than their right-handed counterparts. In line with the findings of Pollet et al. [17] and Dochtermann et al. [18], who did not find a relationship between handedness and winning in the UFC when comparing the outcome of individual fights, no relationship was found between handedness and goalscoring from the 7 m line. However, both Loffing and Hagemann [25] and Richardson and Gilman [3] found evidence for left-handed superiority by studying the population using a different methodology, where the career record of fighters was compared instead of the outcome of any individual fight.

While negative frequency-dependent effects may explain the results of this study, an alternative theory suggesting that left-handers possess innate abilities such as increased neural processing speed and beneficial hormonal configuration that give them advantages in sports has been proposed [4,26]. While the lack of evidence pointing to left-handed advantages in non-invasive sports indicates limited evidence for this theory [4], mixed and inconsistent results from previous studies that only partially confirm the existence of negative frequency-dependent effects leave some researchers suggesting that innate abilities may at the least play some role e.g., [17,27]. This perspective becomes even more relevant in light of evidence that the advantages of negative frequency-dependency can be ameliorated through training [28]. Even though a goalkeeper may have only trained with and played against a handful of left-handed players during their early years in the sport, their constant exposure to left-handed players, and their specific training against left-handed shooters, at the top level should result in increased perceptual familiarity.

Due to the tactical advantages being left-handed offers handball players, left-handed players have an edge when it comes to selection throughout their career (e.g., receiving more playing time, getting into academies, and being more likely to receive a call-up to regional and national training groups and teams [9,29]). However, that does not explain why left-handers are so overrepresented at the 7 m line. If anything, left-handers should be underrepresented when it comes to 7 m throws, at least from a purely logical perspective. The tactical advantages that favor the selection of left-handers are wiped away at the 7 m line, and the right-handed players on the roster, most of whom have had to outperform a greater number of potential candidates to qualify for a spot on the team, would be expected to have superior abilities [30]. However, Loffing et al. [7] found that left-handed players are routinely overrepresented among the top scorers at international tournaments as well, indicating that the advantages of being left-handed in handball may exceed the 7 m line. This may be due to them getting more playing time throughout their career, as left-handers are likely to have less competition for minutes on the field.

The number of 7 m shots any given 7 m shooter took across the four championships varied greatly (from 1 to 64). This is natural as some players took part in all four championships while others only took part in one. Additionally, most teams have a designated first-choice 7 m shooter that takes most of their 7 m shots. When that first-choice option misses a 7 m shot, most teams opt for their second-choice shooter to take the next 7 m shot. It is important to note that the second- and third-choice 7 m shooters are usually highly proficient from the 7 m line, and most likely the first choice at their club team. Any 7 m shot at this level will therefore be taken by a 7 m specialist.

While most studies on negative frequency-dependent advantage from the 7 m spot have been simulated, the current study is based on actual 7 m shots at the world-class level and therefore has high ecological validity. Even though important preliminary knowledge can be gained from controlled simulated experiments, findings have varied depending on the degree of ecological validity [31]. Although the data in the current study are reliable, have high ecological validity, and adequate statistical power, the study design and limited available information related to each 7 m shot constrained which possible covariates could be accounted for. Variables such as shot speed, number of feints, and the time from the referee's whistle to the execution of the shot could therefore not be included in the study.

Future research should examine whether these trends apply to the same degree at the amateur level, because previous research from other sporting contexts has found the degree of negative frequency-dependent advantage to vary depending on proficiency level [17,32]. Slight modifications to the methodology would also be recommended, as analyzing the individual shooters' records as opposed to analyzing every single 7 m shot as a data point could lead to different results [3]. This would require a substantially larger dataset. Examining whether these trends apply to female players would also be of interest, as previous studies have indicated that both left-handedness and lateral biases may be less pronounced among female players [1,5,32]. Finally, future research should address whether left-handers have a scoring advantage during regular gameplay.

5. Conclusions

While the results of this study did not indicate that left-handers were more likely to score from any single 7 m shot, left-handers were greatly overrepresented among the designated shooters compared to both the population as a whole and the proportion of left-handers on any given team. The implication here is that handedness does not appear to play a role in the outcome of 7 m shots at the world-class level, but that handedness does play a role in players becoming world-class 7 m shooters.

Supplementary Materials: The following supporting information can be downloaded at: https://www.mdpi.com/article/10.3390/sym14102163/s1, Supplemental File S1: An overview of handedness at the 106-2022 European championships; Supplemental File S2: An overview of handedness in 7-meter shootouts at the world-class level.

Author Contributions: Conceptualization, A.L., A.I., S.T. and T.H.; methodology, A.I.; formal analysis, A.I.; data curation, A.L.; writing—original draft preparation, A.L.; writing—review and editing, A.L., A.I., S.T. and T.H. All authors have read and agreed to the published version of the manuscript.

Funding: This research received no external funding.

Data Availability Statement: Data is openly available at https://doi.org/10.6084/m9.figshare.20421582.v3 (accessed on 4 October 2022).

Conflicts of Interest: The authors declare no conflict of interest.

References

1. Papadatou-Pastou, M.; Ntolka, E.; Schmitz, J.; Martin, M.; Munafò, M.R.; Ocklenburg, S.; Paracchini, S. Human handedness: A meta-analysis. *Psychol. Bull.* **2020**, *146*, 481. [CrossRef] [PubMed]
2. Llaurens, V.; Raymond, M.; Faurie, C. Why are some people left-handed? An evolutionary perspective. *Philos. Trans. R. Soc. B Biol. Sci.* **2009**, *364*, 881–894. [CrossRef]
3. Richardson, T.; Gilman, R.T. Left-handedness is associated with greater fighting success in humans. *Sci. Rep.* **2019**, *9*, 15402. [CrossRef]
4. Loffing, F.; Hagemann, N. Performance differences between left-and right-sided athletes in one-on-one interactive sports. In *Laterality in Sports*; Loffing, F., Hagemann, N., Strauss, B., Macmahon, C., Eds.; Elsevier: Amsterdam, The Netherlands, 2016; pp. 249–277. [CrossRef]
5. Connor, J.D.; Mann, D.L.; Gomez, M.-A.; Leicht, A.S.; Doma, K. Performance advantages of left-handed cricket batting talent. *Front. Psychol.* **2020**, *11*, 1654. [CrossRef]
6. Raymond, M.; Pontier, D.; Dufour, A.-B.; Møller, A.P. Frequency-dependent maintenance of left handedness in humans. *Proc. R. Soc. Lond. Ser. B Biol. Sci.* **1996**, *263*, 1627–1633. [CrossRef]
7. Loffing, F.; Sölter, F.; Hagemann, N.; Strauss, B. Accuracy of outcome anticipation, but not gaze behavior, differs against left-and right-handed penalties in team-handball goalkeeping. *Front. Psychol.* **2015**, *6*, 1820. [CrossRef]

8. Lucafò, C.; Marzoli, D.; Zdybek, P.; Malatesta, G.; Smerilli, F.; Ferrara, C.; Tommasi, L. The bias toward the right side of others is stronger for hands than for feet. *Symmetry* **2021**, *13*, 146. [CrossRef]
9. Karcher, C.; Buchheit, M. Anthropometric and physical performance requirements to be selected in elite handball academies: Is being left-handed an advantage. *SPSR* **2017**, *9*, 1–2.
10. Schorer, J.; Cobley, S.; Büsch, D.; Bräutigam, H.; Baker, J. Influences of competition level, gender, player nationality, career stage and playing position on relative age effects. *Scand. J. Med. Sci. Sport.* **2009**, *19*, 720–730. [CrossRef]
11. Saavedra, J.M.; Þorgeirsson, S.; Kristjánsdóttir, H.; Chang, M.; Halldórsson, K. Handball game-related statistics in men at Olympic Games (2004–2016): Differences and discriminatory power. *Retos Nuevas Tend. En Educ. Física Deporte Y Recreación* **2017**, *32*, 260–263.
12. Lobinger, B.; Büsch, D.; Werner, K.; Pabst, J.; Gail, S.; Sichelscmidt, P. Erfolgsrelevante Aktionsmuster von Torhütern beim Siebenmeterwurf im Spitzenhandball [Analysis of action patterns of goalkeepers in 7-meter-throws in top-level handball]. *Z. Für Sportpsychol.* **2014**, *21*, 74–85. [CrossRef]
13. Schorer, J.; Panten, J.; Neugebauer, J.; Loffing, F. Perceptual expertise in handball. In *Handball Sports Medicine;* Laver, L., Landreau, P., Seil, R., Popovic, N., Eds.; Springer: Berlin/Heidelberg, Germany, 2018; pp. 597–614. [CrossRef]
14. Cocić, D.; Vaci, N.; Prieger, R.; Bilalić, M. Reading the future from body movements—Anticipation in handball. *J. Mot. Behav.* **2021**, *53*, 483–498. [CrossRef] [PubMed]
15. Loffing, F.; Hagemann, N. Skill differences in visual anticipation of type of throw in team-handball penalties. *Psychol. Sport Exerc.* **2014**, *15*, 260–267. [CrossRef]
16. Hagemann, N. The advantage of being left-handed in interactive sports. *Atten. Percept. Psychophys.* **2009**, *71*, 1641–1648. [CrossRef] [PubMed]
17. Pollet, T.V.; Stulp, G.; Groothuis, T.G. Born to win? Testing the fighting hypothesis in realistic fights: Left-handedness in the Ultimate Fighting Championship. *Anim. Behav.* **2013**, *86*, 839–843. [CrossRef]
18. Dochtermann, N.A.; Gienger, C.; Zappettini, S. Born to win? Maybe, but perhaps only against inferior competition. *Anim. Behav.* **2014**, *96*, e1–e3. [CrossRef]
19. Muthén, L.K.; Muthén, B.O. How to use a Monte Carlo study to decide on sample size and determine power. *Struct. Equ. Model.* **2002**, *9*, 599–620. [CrossRef]
20. Wagenmakers, E.-J.; Marsman, M.; Jamil, T.; Ly, A.; Verhagen, J.; Love, J.; Selker, R.; Gronau, Q.F.; Šmíra, M.; Epskamp, S. Bayesian inference for psychology. Part I: Theoretical advantages and practical ramifications. *Psychon. Bull. Rev.* **2018**, *25*, 35–57. [CrossRef] [PubMed]
21. Muthén, B.; Asparouhov, T. Bayesian structural equation modeling: A more flexible representation of substantive theory. *Psychol. Methods* **2012**, *17*, 313–335. [CrossRef] [PubMed]
22. Zyphur, M.J.; Oswald, F.L. Bayesian estimation and inference: A user's guide. *J. Manag.* **2015**, *41*, 390–420. [CrossRef]
23. Harris, L.J. In fencing, are left-handers trouble for right-handers? What fencing masters said in the past and what scientists say today. In *Laterality in Sports;* Loffing, F., Hagemann, N., Srauss, B., Macmahon, C., Eds.; Elsevier: Amsterdam, The Netherlands, 2016; pp. 31–64.
24. Brooks, R.; Bussiere, L.F.; Jennions, M.D.; Hunt, J. Sinister strategies succeed at the cricket World Cup. *Proc. R. Soc. Lond. Ser. B Biol. Sci.* **2004**, *271*, S64–S66. [CrossRef] [PubMed]
25. Loffing, F.; Hagemann, N. Pushing through evolution? Incidence and fight records of left-oriented fighters in professional boxing history. *Laterality Asymmetries Body Brain Cogn.* **2015**, *20*, 270–286. [CrossRef]
26. Grouios, G. Motoric dominance and sporting excellence: Training versus heredity. *Percept. Mot. Ski.* **2004**, *98*, 53–66. [CrossRef]
27. Witkowski, M.; Tomczak, M.; Karpowicz, K.; Solnik, S.; Przybyla, A. Effects of fencing training on motor performance and asymmetry vary with handedness. *J. Mot. Behav.* **2020**, *52*, 50–57. [CrossRef] [PubMed]
28. Schorer, J.; Loffing, F.; Hagemann, N.; Baker, J. Human handedness in interactive situations: Negative perceptual frequency effects can be reversed! *J. Sport. Sci.* **2012**, *30*, 507–513. [CrossRef]
29. Baker, J.; Kungl, A.-M.; Pabst, J.; Strauss, B.; Buesch, D.; Schorer, J. Your fate is in your hands? Handedness, digit ratio (2D:4D), and selection to a national talent development system. *Laterality Asymmetries Body Brain Cogn.* **2013**, *18*, 710–718. [CrossRef]
30. Gallup, A.C.; Sleicher, E. Left-hand advantage and the right-sided selection hypothesis. *Evol. Behav. Sci.* **2021**, *15*, 184–190. [CrossRef]
31. Huesmann, K.; Loffing, F.; Büsch, D.; Schorer, J.; Hagemann, N. Varying degrees of perception-action coupling and anticipation in handball goalkeeping. *J. Mot. Behav.* **2022**, *54*, 391–400. [CrossRef] [PubMed]
32. Loffing, F.; Hagemann, N.; Strauss, B. Left-handedness in professional and amateur tennis. *PLoS ONE* **2012**, *7*, e49325. [CrossRef] [PubMed]

MDPI AG
Grosspeteranlage 5
4052 Basel
Switzerland
Tel.: +41 61 683 77 34

Symmetry Editorial Office
E-mail: symmetry@mdpi.com
www.mdpi.com/journal/symmetry